Protecting Your Privacy in a Data-Driven World

ASA-CRC Series on
STATISTICAL REASONING IN SCIENCE AND SOCIETY

SERIES EDITORS
Nicholas Fisher, University of Sydney, Australia
Nicholas Horton, Amherst College, MA, USA
Regina Nuzzo, Gallaudet University, Washington, DC, USA
David J Spiegelhalter, University of Cambridge, UK

PUBLISHED TITLES

For more information about this series, please visit: https://www.crcpress.com/go/asacrc

Protecting Your Privacy in a Data-Driven World

Claire McKay Bowen

CRC Press
Taylor & Francis Group
Boca Raton London New York

CRC Press is an imprint of the
Taylor & Francis Group, an **informa** business
A CHAPMAN & HALL BOOK

First edition published 2022
by CRC Press
6000 Broken Sound Parkway NW, Suite 300, Boca Raton, FL 33487-2742

and by CRC Press
2 Park Square, Milton Park, Abingdon, Oxon, OX14 4RN

© 2022 Taylor & Francis Group, LLC

CRC Press is an imprint of Taylor & Francis Group, LLC

Reasonable efforts have been made to publish reliable data and information, but the author and publisher cannot assume responsibility for the validity of all materials or the consequences of their use. The authors and publishers have attempted to trace the copyright holders of all material reproduced in this publication and apologize to copyright holders if permission to publish in this form has not been obtained. If any copyright material has not been acknowledged please write and let us know so we may rectify in any future reprint.

Except as permitted under U.S. Copyright Law, no part of this book may be reprinted, reproduced, transmitted, or utilized in any form by any electronic, mechanical, or other means, now known or hereafter invented, including photocopying, microfilming, and recording, or in any information storage or retrieval system, without written permission from the publishers.

For permission to photocopy or use material electronically from this work, access www.copyright.com or contact the Copyright Clearance Center, Inc. (CCC), 222 Rosewood Drive, Danvers, MA 01923, 978-750-8400. For works that are not available on CCC please contact mpkbookspermissions@tandf.co.uk

Trademark notice: Product or corporate names may be trademarks or registered trademarks and are used only for identification and explanation without intent to infringe.

ISBN: 978-0-367-64077-4 (hbk)
ISBN: 978-0-367-64074-3 (pbk)
ISBN: 978-1-003-12204-3 (ebk)

DOI: 10.1201/9781003122043

Publisher's note: This book has been prepared from camera-ready copy provided by the authors.

Contents

Preface

When I tell people I research data privacy, most assume I specialize in cybersecurity. Perhaps to their disappointment, I correct them that when I use the term "data privacy," I am referencing safely releasing confidential data publicly while preserving the privacy of those who appear in the data. Although this type of data privacy may not seem as exciting as cybersecurity, *it affects every person's life through major public policy decisions in the United States.* Now, you might be wondering, "How is this part of data privacy so important?"

As I write this book, the world is experiencing a global pandemic that has caused severe economic and health public policy issues in most countries, including the United States. If researchers and public policymakers had access to tax and health data, they could better target and coordinate stimulus relief programs to help all American residents. But access to these data could come at a steep privacy cost. For tax data, federal agencies, namely the Internal Revenue Service, cannot release tax data in its raw form because it contains a lot of confidential information. For health data, people have to agree to the collection of their personal, demographic information and real-time movement through smartphone location sharing. In the wrong hands, these data could empower stalkers, endanger children, or facilitate identity fraud.

So, how do academic institutions, federal agencies, and other entities provide access to invaluable data for important public policy decisions while ensuring the privacy of those who are in the data?

Protecting Your Privacy in a Data-Driven World will answer this question and provide several other examples in the context of public policy decisions in the United States. Specifically, the book will address data privacy's importance (Chapter 1), the history and evolution of privacy protection (Chapter 2), the ways privacy experts develop privacy preserving methods (Chapters 3 and 4), these methods' limitations (Chapter 5), the privacy laws governing and protecting people's information (Chapter 6), and other important issues we as society must consider to keep advancing the field of data privacy (Chapter 7).

Your next question may be, "Will I be able to understand this book?" As I wrote the Chapter 1, I decided the intended audience includes anyone interested in learning more about this area of data privacy without a mathematics background. Some of these individuals may include public data users, people working within the state and federal government who are not as familiar with data privacy preserving methods, and public policymakers who want and need to learn more about these methods to make informed policy decisions.

I decided I would want my family to read and understand this book. I am a first-generation college student, and many of my extended family members did not complete college. I had them in mind as my target audience to ensure I explain the technical concepts clearly without complex math. I also want to encourage more people to become interested, and potentially pursue, data privacy and confidentiality as a research topic. There are very few data privacy and confidentiality researchers, especially within statistics, compared with the vast landscape of problems and challenges in the field. I often joke that all the statisticians in this research area can fit in a closet (albeit a larger one). With these motivations in mind, I want to help you better understand the field, without feeling as intimidated or as lost as I was when I first started as a student.

I would be remiss for not mentioning several people who contributed to this book: Len Burman, who reviewed my initial book proposal draft and encouraged me to write a book in the first place. My Telegram "brain trust," who critiqued my book proposal and provided general feedback. Special thanks in particular to Beth Ann Brown, Carl Hunter, Donna Marion, Suzie Neidhart, and Peter Neidhart for their detailed feedback and suggestions that greatly improved the book. Shout-outs to Alexandra Tammaro for helping with extra wordsmithing, and Dallan Duffin for altering the *A Sunday on La Grande Jatte* images for me in Chapter 3. Thank you also to the Carr family, who brainstormed book titles with me as we drank copious amounts of wine and beer for inspiration.

Thank you to my reviewers, Evercita Eugenio, Amy O'Hara, and Stephanie Shipp. Their encouraging comments, constructive feedback, and invaluable insights made the book *much better*.

Of course, I thank my editor, John Kimmel, for keeping me on track and advising me throughout the process. For instance, telling me to rewrite the preface, because I had no clue how to write a preface and wrote a terrible first version. Or reminding me, "Your instinct is to be comprehensive. Whether this is a virtue depends on the audience." Also, cheers to your retirement! Hopefully, my book was not the cause for it...

Finally, I thank and dedicate this book to my spouse, Mack Bowen, who continues to support me both personally and professionally, such as waking up at 3 a.m. to watch me complete my first full-distance triathlon (140.6 miles) and taunting me about the cheeseburger and beer he enjoyed during it.

Author

Dr. Claire McKay Bowen is the Lead Data Scientist for Privacy
and Data Security at the Urban Institute. Her research focuses on
developing and assessing the quality of differentially private data
synthesis methods and science communication. She holds a BS in
mathematics and physics from Idaho State University and an MS
and PhD in statistics from the University of Notre Dame. After
completing her PhD, she worked at Los Alamos National Laboratory, where she investigated cosmic ray effects on supercomputers.

In 2021, the Committee of Presidents of Statistical Societies
identified her as an emerging leader in statistics for her "contribu-
tions to the development and broad dissemination of Statistics and
Data Science methods and concepts, particularly in the emerging
field of Data Privacy, and for leadership of technical initiatives,
professional development activities, and educational programs."

CHAPTER 1

Why Is Data Privacy Important?

O N MARCH 25, 2018, Mark Zuckerberg, CEO of Facebook, published an apology letter through several news outlets about Facebook's role in the Cambridge Analytica Scandal [43]. The incident involved Cambridge Analytica, a British political consulting company, collecting millions of Facebook users' personal data without their consent. They gathered the data by using an app that contained a series of questions to build psychological profiles of each person. Cambridge Analytica then used these profiles for political advertising for the 2016 United States Presidential Campaign and the Brexit referendum.

People did not discover this massive misuse of personal information until a former Cambridge Analytica employee, Christopher Wylie, revealed it to the news and media outlets, the *Observer* and *The Guardian* [8]. The public immediately demanded data privacy reform and pushed for boycotting Facebook, where #DeleteFacebook trended on Twitter. This negative public outcry forced Mark Zuckerberg to testify in front of the United States of America Congress.

Meanwhile at the University of Notre Dame in South Bend, Indiana, and two days after Zuckerberg's initial apology statement, I defended my dissertation titled *Data Privacy via Integration of Differential Privacy and Data Synthesis*. After my concluding remarks, I asked the audience if they had any questions. A few raised their hands and the first person I picked asked how will the Facebook—Cambridge Analytica Scandal affect my research.

DOI: 10.1201/9781003122043-1

1

A few others, who raised their hands, nodded in agreement. They shared the same question about the scandal.

Looking back, the audience probably wanted to know how my current and future research plans would change given Facebook's "breach of trust" in how they handled the scandal. Perhaps to their surprise, I bluntly responded, "The scandal will make it easier to pitch my research to funders."

1.1 WHAT IS DATA PRIVACY?

Although the scandal heightened the public's concern about how personal data should be collected and used, the field of data privacy and confidentiality has existed for decades. The Facebook— Cambridge Analytica Scandal did not reveal any new data privacy challenges. Instead, it brought to light the privacy and public policy issues of how private companies gather and disseminate the public's personal information.

In particular, modern computing and technology has made the collection, storage, and analysis of larger and more complex data possible. However, this computational advantage comes at a heavy cost to privacy. Malicious actors can more easily reconstruct databases and identify individuals from supposedly anonymized data with modern computing by linking these private records to public databases.

One of the classic examples of this kind of data attack, known as a *record linkage* attack, is the re-identification of the then Massachusetts Governor William Weld. In 1997, Governor Weld announced the public release of the Massachusetts Group Insurance Commission data for researchers to analyze ways to improve healthcare and control rising associated costs. Governor Weld assured the public that staff secured the data by deleting personally identifiable information prior to release.

A few days later, Dr. Latanya Sweeney, then a Massachusetts Institute of Technology graduate student—now a Harvard Professor of Computer Science, mailed to Weld's office his personal medical information. Sweeney purchased voter data for twenty dollars and used that data to link Weld's birth date, gender, and zip code to his health records in the Group Insurance Commission data.

Although this example occurred over two decades ago, it demonstrates how easily we can identify people with other data sources. Additionally, Sweeney had the limited computational power of the late 90s. To put it in perspective, the smart phones of the early 2020s have more computational power than the average desktop an American owned in the early 2010s.

With this and many more data privacy challenges, researchers in the field constantly develop and update disclosure control approaches (or methods of data privacy and confidentiality to safely release data publicly) to prevent privacy violations, such as re-identification. Again, these problems have already existed for decades, which is why the Facebook—Cambridge Analytica Scandal did not change or veer my future research path. It provided more motivation and reasons why we should all be wary of how our personal information is being collected and used, even "behind closed doors."

For the remainder of this book, I will use the term *data privacy* instead of stating *data privacy and confidentiality* to keep the phrase more concise. However, these terms have a subtle difference. Data privacy refers to the individuals' sensitive information that they do not want exposed, whereas data confidentiality refers to how we protect peoples' information, such as limiting access to the sensitive data. When I use the term data privacy in this book, I am referring to how researchers gain access to confidential data while protecting the privacy of those who are in the data.

1.2   WHY SHOULD *ANYONE* CARE?

When using social media platforms, some people do not care if others know that they "thumbs up" cat videos, share details of their recent vacation, or list basic personal information, such as their birthday and work info. For instance, if you go to my social media, you will see I am married, I live in New Mexico, I love cats and dogs, and I have an unhealthy obsession with racing in triathlons. Many people share this information because we are either unconcerned about the information being public or unaware of the potential data privacy attacks.

Let us imagine another type of data being collected. What if someone collected location data from our smart phones. Similar to the Group Insurance Commission data, that person wants to remove the personally identifiable information and apply other disclosure control methods, such as grouping some information into broader categories. They decide to only leave a few demographic details, date, time, and location (i.e., latitude and longitude coordinates) before publicly releasing the data.

Suppose we analyze the data and follow the movements of one person. We see that during standard working hours, this person is at the Redmond Microsoft campus. We observe this trend over several months, increasing the likelihood the person works at Microsoft. We then see a change. The person briefly went to the Amazon campus one morning before resuming their daily commute to Microsoft. A month later this person starts regularly traveling to Amazon.

This scenario actually happened. Reporters from *The New York Times* wrote about how they used cell phone location data to identify the person as Ben Broili, who switched jobs from Microsoft to become the manager for Amazon Prime Air, the drone delivery service [60]. Even after learning this, some might think that knowing where someone works is not revealing too much information. For example, anyone can find where I work on LinkedIn, Twitter, Facebook, and my personal website. Large data companies, such as Foursquare, also use location data to evaluate other companies' phone advertisement performance or understand our purchasing behaviors. This all seems harmless.

But, the United States currently has no data privacy laws that regulates how consumer data can be collected, used, sold, shared, and stored. Reporters from Motherboard—Vice discovered that bounty hunters purchased location data from data brokers to find their target's cell phone numbers in real time [11]. Revisiting the Ben Broili example, what if a company decides to prematurely fire employees who might be interviewing elsewhere? In another scenario, a malicious person uses location data to identify a family and study their routine. The person easily learns where the children go to school and when the children are home alone. A car insurance company links the records of their internal data to the cell phone location data. The company then determines which of their current or future clients are likely to have a health condition based on how frequently they visit the doctor and raises insurance premiums for those individuals.

Given these possibilities, many might demand that such data should not be collected. The risks are too high. But, what if we used the data for the benefit of society?

In combating the coronavirus pandemic, most proposed contact tracing strategies require accurate location data to be effective [50]. Researchers analyze mobile phone data to provide various emergency management scenarios, such as terrorist attacks or earthquakes [62]. The 500 Cities Project created a large United

States dataset that "contain[ed] estimates for twenty-seven indicators of adult chronic disease, unhealthy behaviors, and preventative care available." This collaborative project with the Centers for Disease Control and Prevention (CDC), the Robert Wood Johnson Foundation, and the CDC Foundation aimed to provide a "groundbreaking resource for establishing baseline conditions, advocating for investments in health, and targeting program resources where they are needed most" [52].

As we think through these examples, we notice some of the tension between personal privacy and the common good. We are forced to ask ourselves, "At what point does the sacrifice to our personal information outweigh the public good?" Revealing too much information places people at risk, such as empowering stalkers to more easily track people. Collecting too little information will restrict our ability to help people, such as implementing contact tracing apps.

1.3   WHY IS BALANCING DATA PRIVACY AND UTILITY HARD?

The answer to this question is complicated, so I will illustrate issues that we need to navigate with myself as an example [5].

Suppose I participated in a census survey at the beginning of 2020. I want to keep my identity private, but my accurate representation in the data is also important. At that time, I was a millennial Asian American woman residing in Washington, DC. Because of the size and racial diversity of Washington, DC, the United States Census Bureau can easily "hide" me in the data while also preserving certain statistical qualities: that there are already many millennial, Asian American women in Washington, DC.

Issues of data privacy quickly arise in areas with much smaller populations. Years ago, I grew up in a very remote town in Idaho called Salmon. This town has a population of 3,096[1] and is the largest town in a county that is roughly the size of Connecticut.[2] "Hiding" me in Salmon, where I was the only Asian American high school student, is much harder. To keep my identity hidden, a data privacy method could either erase me from the picture of

---

[1]United States Census Bureau population estimate for 2019.
[2]Salmon is in the Lemhi County, which is 4,569 square miles. Connecticut is 4,845 square miles.

Salmon, Idaho or inaccurately add more Asian American women in the small town.

This simple example highlights how data privacy and data utility (or usefulness) oppose one another. We cannot have all the privacy and all of the utility. The general data privacy community that consists of data stewards, data participants, data practitioners, researchers, and public policymakers must weigh the costs and benefits in order to balance the two sides. But, the challenge of balancing them becomes even more difficult in other scenarios or data structures, which I will outline in Chapter 5.

1.4 WHY IS THERE INEQUALITY IN PRIVACY?

These conversations about data privacy and data utility usually overlook how the personal privacy loss is not distributed equally in society. Underrepresented individuals, most notably racial minorities and socially economically disadvantaged individuals, experience higher privacy insecurity and are at a higher risk to their data privacy [39]. Yet, these individuals are more likely to suffer worse public health outcomes [16].

What do I mean by this? Let us walk through another example. Fewer Hispanic immigrant families reside in the Midwestern part of the United States than Caucasian families. If we released data on family composition with gender and race for this geographic region, people accessing the data could easily identify Hispanic immigrant families in rural areas, violating the Hispanic families' data privacy. On the other hand, we could protect the Hispanic families' data by not releasing their information or changing the data values. However, removing or altering only underrepresented groups' data may result in those individuals not reaping the same benefits from that data being released for research or other public health purposes.

In recent years, the number of Hispanics working in meat-processing plants has significantly increased in the United States. Although they are the majority in some cities, Hispanic families are still the minority in rural counties. To make matters worse, most healthcare data contain no ethnic or racial information. The lack of data creates a cultural disconnect between the people using the data, who often assume incorrect racial and ethnic breakdown, and the populations that need the most help. This leads to health officials providing insufficient warnings and health procedures on how to prevent and combat major health crises, creating more discrimination issues. In 2020, for example, we saw a large increase in COVID-19-related deaths among Latino meat processing plant workers [18].

That same year in May, the Navajo Nation surpassed New York City for the highest per capita coronavirus infection rate [55]. When the United States first started to shelter-in-place, many of the health advisories included washing your hands for 20 seconds. However, the Navajo Nation, where over a third of the population lacks running water, could not apply this healthcare advice [22]. Other Native American populations suffered from similar problems, such as the pueblos of New Mexico, which remained closed to the public until early 2021 when the state administered vaccines.

To make matters worse, researchers expect that people who test positive for the coronavirus are at a higher risk of being identifiable in released data based on past medical data privacy incidents. The data privacy community must address these problems or else create worse outcomes for underrepresented individuals in both health and privacy.

1.5   WHAT WILL BE COVERED IN THIS BOOK?

As you noticed, I wrote this book as a progression of questions that lead to one another because I envisioned you as myself when I was a student. I thirsted for the answers to my many questions, but created more questions instead of solving the ones I originally posed. This thought process led me to write the book format that outlines the history of data privacy, the past and current approaches to preserving data privacy, the reasons why some data are more difficult to protect than others, the need for better data privacy laws, and the future data privacy challenges I believe we should focus on. I will also continue to highlight the data privacy inequality for underrepresented individuals in various scenarios.

Throughout the book, I will refer to various groups within the data privacy community. This community consists of privacy experts, researchers (e.g., social scientists, economists, and demographers), data stewards, data practitioners, and public policymakers. Privacy experts or privacy researchers are individuals who specialize in developing data privacy methods. Researchers are experts in another field of study, such as economics, and are connected to the data privacy community because they need access the confidential data. Data stewards, data curators, or data maintainers are individuals or institutions that are responsible for the collection and storage of the confidential data. Data practitioners or data users are analysts and users of the publicly released version of the confidential data. They also often help the data stewards with applying data privacy methods on the confidential data from the perspective of those who will ultimately use the data. With the book's focus on American public policy and social issues, the public policymakers are the various stakeholders in the United States of America who will ultimately make the data privacy decisions the country must follow.

Data privacy affects our lives everyday and becomes a greater burden on underrepresented groups. As a society, we must take a more active role in deciding how we balance protecting our data while ensuring data quality for the public good.

CHAPTER 2

How Did Data Privacy Change Over Time?

W HEN I STARTED WRITING THIS BOOK, the United States Census Bureau conducted the 2020 Census. At the time, you might have read or heard news stories about the 2020 Census because the data affects how the United States apportion the 435 seats for the United States House of Representatives, redistrict voting lines, plan for natural disasters, and allocate the $1.5 trillion budget, among many other things. During that year, we experienced the COVID-19 pandemic and related stay-at-home orders, natural disasters, changed household dynamics, eviction freezes, displacement of college students, challenges to the trust in government and reliability of institutions, and national social upheaval. The coronavirus also delayed the United States Census Bureau operations, and the Trump administration truncated census fieldwork two weeks earlier than initially scheduled. Each of these factors will have lasting political and economic ramifications.

Unfortunately, the Census Bureau is no stranger to these types of events. It has a long and tumultuous record of political and ethical controversies, and now a pandemic that all ties into our need for safely releasing data publicly and securely accessing confidential data. For these reasons, I will use the United States Census Bureau as an example throughout this chapter to explain the history and evolution of the United States' data privacy protection. I will cover when the United States started protecting personal information and what events shaped how the Census Bureau and other federal agencies collect, use, and share personal data. I will then

DOI: 10.1201/9781003122043-2

end the chapter on what federal agencies are doing now, as of the publication of the book, to improve our access to confidential data.

For readers interested in the United States Census Bureau's history, I recommend checking out the Census Bureau's official website on its history and a book by Margo J. Anderson called, *The American Census: A social history* [25, 2].

2.1 HOW DID DATA PRIVACY BEGIN FOR THE UNITED STATES CENSUS BUREAU?

In 1790, a year after George Washington was inaugurated, the United States marshals conducted the first census as mandated by the United States Constitution.[1] The Founding Fathers of the United States included the census in the United States Constitution to ensure that every community has the correct number of government representatives. As a form of quality control, Congress required the marshals post the census lists in town squares for the community to review. This process allowed the public to verify if they were counted properly. If not, the residents could correct the mistake. As someone who grew up in a small town, I understand how this would be effective in ensuring an accurate count. Everyone in Salmon knew where everyone else lived, who their family members are, what pets they have (if any), and what they drove.

While this procedure guaranteed data quality, there was no privacy. This changed when the marshals started collecting manufacturing business data in 1810. Unlike the general population census, the manufacturing business response was voluntary. Congress thought that individuals involved in the manufacturing industry would willingly answer the questions given the demand for better manufacturing data. The responses were dismal.

The marshals filed the statistical tables for the manufacturing data with the United States District Court clerks instead of publicly posting them in town squares. However, most people raised confidentiality concerns due to not trusting their response would be properly protected. This reaction resulted in extremely poor manufacturing data coverage and quality in the 1810 and 1820 censuses, which prompted Congress to remove the manufacturing questions for the 1830 Census. Congress then restored the manufacturing questions for the 1840 Census after they assured the public

---

[1]The United States census is mandated under Article I Section 2 of the United States Constitution.

that no individual or business names would appear in the official statistics tables, understanding the need for confidentiality.[2]

In the 1850 Census, Congress finally ordered the United States marshals to stop posting results in public locations. Years later, they established a Census Office to ensure the quality and privacy of the 1870 Census. Congress also enacted a new privacy law for the following census in 1880. The law imposed heavy fines if Census Office staff members, such as the census takers and clerks, broke their oath of secrecy.

However, all these changes focused on *direct disclosure risks* and not *indirect disclosure risks*. For example, a staff member releasing specific records on businesses and individuals is a *direct disclosure risk*. An *indirect disclosure risk* would be someone isolating a particular manufacturing business based on knowing that a county only has two manufacturing businesses. This indirect disclosure risk attack is a record linkage attack that was described in the previous chapter (albeit without using computers).

The major turning point for the census came on July 1, 1902, where the Census Office became a permanent entity now known as the United States Census Bureau. Then, on July 2, 1909, Congress approved the 1910 Census Act (see Figure 2.1). This act provided the Census Bureau additional funds to increase permanent staff, changed Census Day from June 1 to April 15, and stated that all employees of the Census Bureau are "prohibited, under heavy penalty, from disclosing any information which may thus come to his knowledge." These penalties include both jail time and hefty fines.

The act also required, for the first time, data privacy protection against indirect disclosure risks for business data but did not protect the general population data. Early attempts at preventing indirect privacy loss entailed the Census Bureau staff looking through data tables. If they found any potentially vulnerable values, they manually suppressed (i.e., removed values) or aggregated results (i.e., combined smaller group of values into larger groups).

Although Congress recently changed Census Day, the Department of Agriculture requested the day be shifted again to January 1 for the 1920 Census. The department thought that farmers would complete more of their harvest and would remember their harvest

---

[2]The United States Census Bureau publishes dozens of tables on various statistics that include School Enrollment, Household Pulse Surveys, and Geographic Mobility.

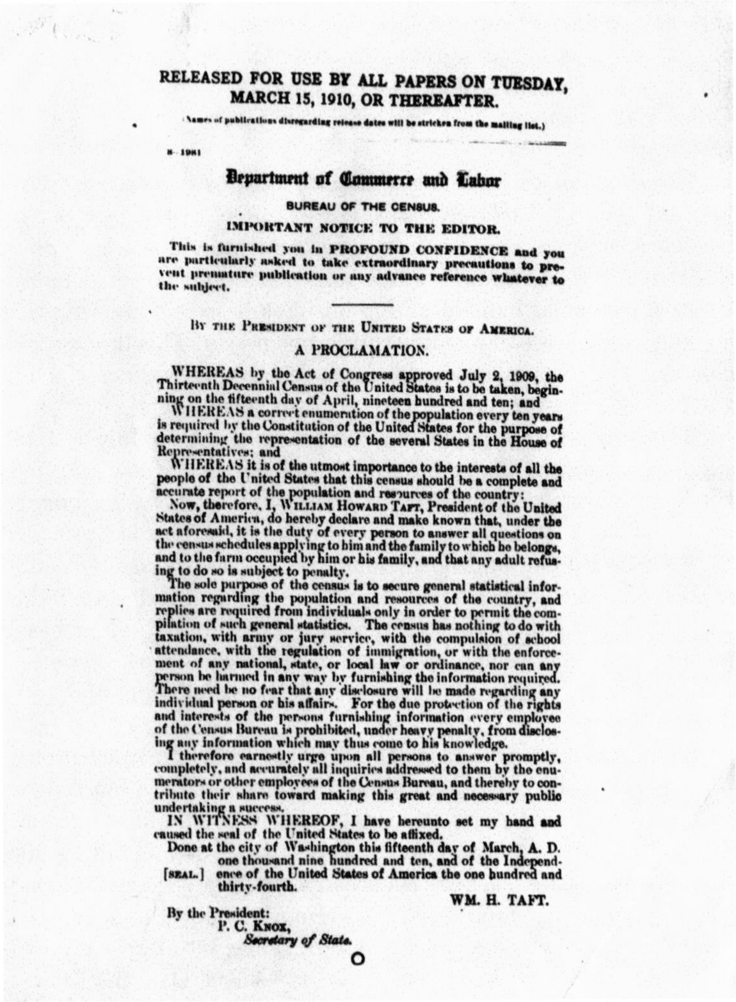

RELEASED FOR USE BY ALL PAPERS ON TUESDAY, MARCH 15, 1910, OR THEREAFTER.

(Names of publications disregarding release dates will be stricken from the mailing list.)

8—1981

Department of Commerce and Labor

BUREAU OF THE CENSUS.

IMPORTANT NOTICE TO THE EDITOR.

This is furnished you in PROFOUND CONFIDENCE and you are particularly asked to take extraordinary precautions to prevent premature publication or any advance reference whatever to the subject.

BY THE PRESIDENT OF THE UNITED STATES OF AMERICA.

A PROCLAMATION.

WHEREAS by the Act of Congress approved July 2, 1909, the Thirteenth Decennial Census of the United States is to be taken, beginning on the fifteenth day of April, nineteen hundred and ten; and

WHEREAS a correct enumeration of the population every ten years is required by the Constitution of the United States for the purpose of determining the representation of the several States in the House of Representatives; and

WHEREAS it is of the utmost importance to the interests of all the people of the United States that this census should be a complete and accurate report of the population and resources of the country:

Now, therefore, I, WILLIAM HOWARD TAFT, President of the United States of America, do hereby declare and make known that, under the act aforesaid, it is the duty of every person to answer all questions on the census schedules applying to him and the family to which he belongs, and to the farm occupied by him or his family, and that any adult refusing to do so is subject to penalty.

The sole purpose of the census is to secure general statistical information regarding the population and resources of the country, and replies are required from individuals only in order to permit the compilation of such general statistics. The census has nothing to do with taxation, with army or jury service, with the compulsion of school attendance, with the regulation of immigration, or with the enforcement of any national, state, or local law or ordinance, nor can any person be harmed in any way by furnishing the information required. There need be no fear that any disclosure will be made regarding any individual person or his affairs. For the due protection of the rights and interests of the persons furnishing information every employee of the Census Bureau is prohibited, under heavy penalty, from disclosing any information which may thus come to his knowledge.

I therefore earnestly urge upon all persons to answer promptly, completely, and accurately all inquiries addressed to them by the enumerators or other employees of the Census Bureau, and thereby to contribute their share toward making this great and necessary public undertaking a success.

IN WITNESS WHEREOF, I have hereunto set my hand and caused the seal of the United States to be affixed.

Done at the city of Washington this fifteenth day of March, A. D. one thousand nine hundred and ten, and of the Independence of the United States of America the one hundred and thirty-fourth.

[SEAL.]

WM. H. TAFT.

By the President:
P. C. KNOX,
    Secretary of State.

Figure 2.1: President William Howard Taft's March 1910 Census Proclamation. Image from census.gov website.

yields better by that date. They also argued that more people would be home in January than in April.

On June 18, 1929, Congress enacted the Fifteenth Census Act[3] that stated the 1930 Census would be "a census of population, agriculture, irrigation, drainage, distribution, unemployment, and mines." With these new statistics, Census Day changed to be April 1 and has been that day since then.

2.2   HOW DID TITLE 13 BECOME LAW?

Now, some might be wondering why there was not a United States law that protected *individuals* against indirect disclosure risks. There is a law, Title 13 United States Code, which requires the United States Census Bureau to "...provide strong protection for the information [the Census] collect[s] from individuals and businesses." In other words, from the Census Bureau web-page:

- ***Private information is never published.*** *It is against the law to disclose or publish any private information that identifies an individual or business such, including names, addresses (including GPS coordinates), Social Security Numbers, and telephone numbers.*

- ***The Census Bureau collects information to produce statistics.*** *Personal information cannot be used against respondents by any government agency or court.*

- ***Census Bureau employees are sworn to protect confidentiality.*** *People sworn to uphold Title 13 are legally required to maintain the confidentiality of your data. Every person with access to your data is sworn for life to protect your information and understands that the penalties for violating this law are applicable for a lifetime.*

- ***Violating the law is a serious federal crime.*** *Anyone who violates this law will face severe penalties, including a federal prison sentence of up to five years, a fine of up to $250,000, or both.*

---

[3]The Fifteenth Census Act is also known as the Reappointment Act of 1929 that permanently capped the number of Congressional seats to 435 for the United States House of Representatives.

Although these tenants may seem obvious, Title 13 had to undergo several censuses, a world war, and multiple questionable use of the census data to exist in its current form.

The idea of Title 13 started when Congress passed the 1929 Census Act, stating the census "shall be used only for the statistical purposes for which [they are] supplied." This act enforced that individuals and businesses cannot be identified in publicly released data. For individual data, only the person and their descendants can access that person's records.

In other words, the 1929 Census Act extended indirect disclosure risk protection to individuals and not just businesses. This change created an interesting consequence for the 1930 Census. The Census Bureau decided to not publish small-area data, because they could not prevent indirect disclosure risk. The challenge of publishing smaller populations persists today and will be discussed further in Chapter 5, Section 5.4.

With the 1929 Census Act, the United States Census Bureau prevented other federal offices from accessing the confidential data. In 1930, the Women's Bureau requested information on all women living in Rochester, New York State, containing their names, addresses, occupations, and employment status. The Census Bureau solicited the United States Attorney General's[4] advice, who determined the data could not be released due to the new act. Several more entities, such as law-enforcement and security agencies, have asked for census data. The Census Bureau denied those requests.

These incidences and many others seemed to indicate a positive progression on protecting the data privacy of individuals and businesses. Congress also restructured and transformed the act into Title 13 United States Code in 1954. But, there were a few obstacles along the way.

One hurdle was the Director of the Census Bureau could disclose any individual census records at their discretion. This interpretation of the code originated during World War I, when the United States government requested access to individual census records. At the time, only the 1910 Census Act law existed, which did not prohibit disclosure of individual records. This means the Census Bureau legally provided data for the United States Department of Justice and local draft boards. The Census Bureau

---

[4]The United States Attorney General is the head of the United States Department of Justice, the chief lawyer of the federal government of the United States, and a member of the Cabinet of the United States.

also sent the United States Provost Marshal General[5] national estimates of men in various age groups to compare registration results.

These actions started the interpretation that the Census Director could release census information at their discretion. In 1917, the United States Provost Marshal General stated

> [The] Director of the Census might, in the exercise of his[6] discretion, furnish to the officials in charge of the execution of the Selective Service Law, information in regard to the names and ages of individuals, as it did not appear that any person would be harmed by the furnishing of such information for the purpose for which it was desired.

While the Census Bureau was conducting the 1920 Census, the United States Department of Justice wanted citizenship information from census officials for the United States Department of Labor's deportation cases. The Solicitor General[7] held the same interpretation as the United States Provost Marshal General, but added the Director of the Census Bureau should consider if the request would interfere with the ongoing census.

These requests highlight a loop hole in the 1920 Census Act—it did not legally prevent the Census Bureau Director from releasing information about individuals. When Title 13 became law, it stated the Census Director could release census data on a person to governors, courts, and other individuals at their discretion. Only in 1976 did Congress amended Title 13 to prevent the Census Bureau Director from granting access to census data.

The next "setback" occurred after Pearl Harbor. The War Department requested the 1940 census data on Japanese Americans at the census tract and block levels (see Figure 2.2 for Census geographic summary levels). The United States Census Bureau never confirmed nor denied if they obliged to the request until a past Census Bureau Director, Vincent Barabba, wrote in a memorandum dated 1980, stating

---

[5] The United States Provost Marshal General is the highest-ranking provost marshal position in the United States Army.

[6] In the entire history of the United States Census Bureau, there has only been two female directors, who served consecutively from 1989–1998.

[7] The Solicitor General is the fourth highest-rank within the United States Department of Justice.

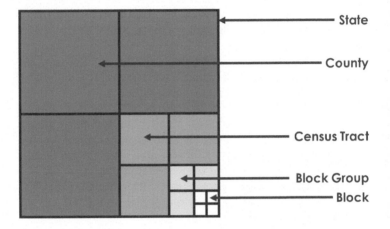

Figure 2.2: An image of Census geographic summary levels.

A Government report, "Japanese Evacuation from the West Coast," notes that the 1940 census data were the single most important source of information used for evacuation and resettlement purposes, and the Bureau prepared a duplicate set of punch cards on Japanese Americans to assist in this effort. Shortly after the United States entered the war late in 1941, the war agencies asked that a Census Bureau statistician be assigned to assist them. Early in 1942 a statistician was transferred to California for this work. He requested and obtained a duplicate set of punch cards from which aggregate data were tabulated on an expedited basis. Virtually all of the data tabulated at this time was eventually published in the 1940 census reports issued in 1943. The 1940 census publications showed separate statistics on Japanese Americans by counties within States. Similar data on racial or ethnic origin had been published from the 1930 census, though in less detail. The extra set of punch cards permitted the 1940 aggregated information to be compiled on a more rapid basis than the normal schedule for published census reports.

Despite the admittance, people still speculated whether the United States Census Bureau violated any of the laws that were enacted at the time. To clear the confusion, Kenneth Prewitt, who was a past Census Bureau Director, wrote an email to all Census Bureau staff on March 24, 2000:

> The historical record is clear that senior Census Bureau staff proactively cooperated with the internment, and that census tabulations were directly implicated in the denial of civil rights to citizens of the United States who happened also to be of Japanese ancestry.
>
> The record is less clear whether the then in effect legal prohibitions against revealing individual data records were violated. On this question, the judicial principle of innocent until proven otherwise should be honored. However, even were it to be conclusively documented that no such violation did occur, this would not and could not excuse the abuse of human rights that resulted from the rapid provision of tabulations designed to identify where Japanese Americans lived and therefore to facilitate and accelerate the forced relocation and denial of civil rights.
>
> I would also like to state clearly that for many years the Census Bureau was less than forthcoming in publicly acknowledging its role in the internment process. Silence was not the worst offense, for there is ample evidence that at various times the Census Bureau has described its role in such manner as to obfuscate its role in internment. Worst yet, some Census Bureau documents would lead the reader to believe that the Census Bureau behaved in a manner as to have actually protected the civil rights of Japanese Americans. This distortion of the historical record is being corrected.
>
> The internment of Japanese Americans was a sad, shameful event in American history, for which President Clinton, on behalf of the entire federal government, has forthrightly apologized. The Census Bureau joins in that apology and acknowledges its role in the internment.
>
> In the post-war period, important safeguards to protect against the misuse of census tabulations have been instituted, notably stronger legal provisions to protect data confidentiality and the Census Bureau's introduction of disclosure avoidance techniques. Adherence to

> *these safeguards preclude a repeat of the 1941/42 be-*
> *havior.*

This tragedy motivated stricter data privacy regulations and the formal development of disclosure control or avoidance techniques with the Census Bureau. It also emphasizes the importance of how anyone who collects, analyzes, and disseminates data has a moral obligation to protect the data to the best of their ability.

Another event that affected federal departments occurred in 1972. The Nixon administration attempted to cover up their break-in of the Democratic National Committee headquarters, known as the Watergate scandal. In light of the scandal, Congress wanted to impede illegal surveillance and investigation of individuals by federal offices. Congress also worried about how much confidential information federal agencies stored on computers, making it easy for someone to retrieve specific personally identifiable information, such as Social Security numbers.

Given these concerns, Congress enacted the Privacy Act of 1974 that stated "No agency shall disclose any record which is contained in a system of records by any means of communication to any person, or to another agency, except pursuant to a written request by, or with the prior written consent of, the individual to whom the record pertains..." In other words, the act established how personally identifiable information should be collected, maintained, used, and disseminated within federal agencies.

These historical events highlight how United States laws changed from one extreme to the next for releasing and sharing confidential administrative data, or data collected by the federal government for a specific service. These stricter privacy laws did not ease until Congress amended Title 13 in the 90s and 2000s. In 1994, Congress allowed the United States Census Bureau to share census address information with state and local governments for completing future censuses and surveys. Then, in 1997, the Section 9a amendment created the Census Monitoring Board to prepare and implement the 2000 Census, permitting the board members to access the confidential census data to carry out their duties. As of the publication of this book, the last amendment occurred in 2002 under the Confidential Information Protection and Statistical Efficiency Act to allow the Census Bureau to share business data to the United States Bureau of Economic Analysis and the United States Bureau of Labor Statistics.

Although I focused on the United States Census Bureau, other federal departments are bound by similar, but different laws. The Confidential Information Protection and Statistical Efficiency Act enforces how the Bureau of Labor Statistics, Bureau of Economic Analysis, the National Center for Health Statistics, and the Census Bureau use survey data. For the Internal Revenue Service, they must follow Title 26 (i.e., Section 6103) on who can access tax data, what are the expectations for data privacy, and the legal ramifications if there are unauthorized disclosures.

For some data, both Title 13 and Title 26 will protect data together. The Census Bureau integrates administrative tax information from the Internal Revenue Service into many of their data products. An example is the Longitudinal Business Database that consists of information on business establishments, firms, and employees for all industries throughout the United States.

2.3   WHAT ARE OTHER UNITED STATES LAWS THAT REGULATE FEDERAL STATISTICS?

Congress established many of the laws we have covered so far several decades ago. This begs the question, "What has the United States done more recently to regulate how federal agencies collect, use, and share data?"

On March 30, 2016, Congress passed the Evidence-Based Policymaking Commission Act of 2016 to create a commission that would "...study and develop a strategy for strengthening government's evidence-building and policymaking efforts" through better use of existing government data. The act detailed the selection of fifteen members[8] from academia, industry, and government. Some of the members included the Director of the Office of Management and Budget,[9] privacy experts, and economic professors. For the remainder of this chapter, I will refer to the Evidence-Based Policymaking Commission as "the Commission" for simplicity.

Over an 18 month period, the Commission conducted thorough fact-finding and deliberation processes before summarizing their findings in a final report called, "The Promise of Evidence-Based

---

[8] Evidence-Based Policymaking Commission Act of 2016 specified the president, the House Speaker, the House Minority Leader, the Senate majority leader, and Senate minority leader each appoint three people for a total of fifteen.

[9] Office of Management and Budget creates the president's budget and oversees the performance of federal agencies.

Policymaking: Report of the Commission on Evidence-Based Policymaking." In their letter to Congress, they stated they "...envision[ed] a future in which rigorous evidence is created efficiently, as a routine part of government operations, and used to construct effective public policy" [46].

While using data more effectively for better public policymaking is great, some might wonder, "Does this mean we will also increase privacy risks?" In Chapter 1, we learned that data privacy and data utility naturally oppose one another. However, during the Commission's investigation, they concluded that:

> *Traditionally, increasing access to confidential data presumed significantly increasing privacy risk. The Commission rejects that idea. The Commission believes there are steps that can be taken to improve data security and privacy protections beyond what exists today, while increasing the production of evidence. Modern technology and statistical methods, combined with transparency and a strong legal framework, create the opportunity to use data for evidence-building in ways that were not possible in the past.*

Given this statement, some might ask, "What are the Commission's suggestions on safely expanding confidential data?" In the report, the Commission made twenty-two recommendations under three themes:

1. **"How the Federal government can provide the infrastructure for secure access to data" or Improving Secure Access to Confidential Data.** Congress has not optimized most United States laws to facilitate data sharing across various federal programs. For instance, if a federal agency wants access to administrative data from another federal agency that is not specifically supported by law (e.g., Title 13 and Title 26), then the agencies must negotiate the terms of accessing the data *every fiscal year*. The negotiation process can take several months, wasting valuable time and money. Some recommendations to improve safe access to confidential data for evidence-building include:

   - establish the National Secure Data Service, a service that will provide secure data access of multiple confidential data while ensuring transparency and privacy

- review and revise laws to allow statistical use of survey and administrative data for federal programs

2. **"The mechanisms to improve privacy protections and transparency about the uses of data for evidence-building" or Modernizing Privacy Protections.** Each federal office has a different standard for what data privacy methods are acceptable to minimize disclosure risks. Many of these offices also use "older" privacy methods that are not effective against modern technology and computing. Some recommendations to modernize privacy protections for evidence-building include:

   - require all federal agencies to conduct a comprehensive risk assessment on their public data that originated from confidential data to improve how the data are protected

   - adopt state-of-the-art privacy-preserving technologies to improve how data are shared, released, and protected

3. **"The institutional capacity to support evidence-building" or Strengthening Federal Capacity.** Federal offices will need to develop evidence-generation programs, provide administrative flexibility, and build workforce to ensure effective interagency collaborations on increasing data access and quality. Some recommendations to strengthen federal capacity for evidence-building include:

   - establish a Chief Evaluation Officer in each federal office to facilitate collaborations across federal agencies

   - ensure sufficient resources to support these activities

Since the release of the report, the United States government has addressed several of the Commission's recommendations. On January 14, 2019, Congress passed the Foundations for Evidence-Based Policymaking Act of 2018, often referred to as the Evidence Act. This bipartisan act covers roughly half of the recommendations the Commission outlined in three parts, which are:

1. **Federal Evidence-Building Activities.** These activities include establishing evaluation officers, designating statistical officials and interagency council on statistical policy, and requiring the heads of each federal agency to create agency evidence-building and evaluation plans. The Evidence Act

also created an Advisory Committee on Data for Evidence-Building to "...review, analyze, and make recommendations on how to promote the use of Federal data for evidence-building."

2. **Open, Public, Electronic, and Necessary (OPEN) Government Data Act.** This new act aims to improve access to public data that can be invaluable for government and private sector functions and innovations. OPEN Government Data Act accomplishes this goal by requiring all federal agencies to make their data open and available by default (if possible), meet certain data standards,[10] and provide accountability through reporting to the Government Affairs Office. In addition, the act established Chief Data Officers at each federal agency to manage the new responsibilities on data management along with internal and external operations and collaborations.

3. **Confidential Information Protection Statistical Efficiency Act of 2018.** The updated act empowered statistical agencies to use more administrative data for statistical purposes and required these agencies to reevaluate the disclosure risks on their data. Overall, the act restructured these privacy laws, reinforced confidential data protections, encouraged collaborations and statistical efficiency, and expanded secure access to other data under the act.

Other progress include a Federal Data Strategy, a ten-year plan to improve the United States government data access, management, and use. The strategy also included several documentations from the Office of Management and Budget on how federal programs should implement the Commission's recommendations.

Although the United States government has made great strides in increasing more evidence-based public policymaking, the federal government has not implemented several other important recommendations. For example, the deployment of the National Secure Data Service that would allow data users to combine multiple, confidential data in a secure environment to increase evidence-building. However, developing the National Secure Data Service is not an easy undertaking and will require extensive planning,

---

[10]Example of the data standards are data must be machine-readable (i.e., data that are in a format that can be easily processed by a computer) and contain metadata (i.e., structural and descriptive information about data).

preparation, and coordination among several federal offices to implement appropriately [23].

Moreover, implementing the Commission's recommendations can create unintended issues. For instance, due to declining survey responses, some federal agencies started using administrative data instead of survey data to create current and new data products. But, some of the laws have not "caught up" to this change.

The updated Confidential Information Protection Statistical Efficiency Act, for example, only specifies that data from individuals and organizations must be protected. Some might ask, "What if the data are provided by another source through the Evidence Act?" In these cases, "legal requirements to protect [administrative] data are not always as straightforward as for survey data," where some of the newly created data are not protected by Confidential Information Protection Statistical Efficiency Act [17].

What does this all mean for American society? The United States government is at various stages of implementing the Commission's recommendations and the Foundations for Evidence-Based Policymaking Act. Furthermore, many public and private institutions in collaboration with federal agencies are working together to address the remaining recommendations, such as the National Secure Data Service. More on that in Chapter 7, Section 7.2.

The United States government will take several years to fully realize the Commission's recommendations and Evidence Act. Once completed, these changes will be transformative in how the United States government will make more data-informed and evidence-based public policy decisions. When this happens, the federal government and society must ethically and safely expand access to confidential data or risk repeating history.

CHAPTER 3

How Do Data Privacy Methods Expand Access to Data?

A S A DOCTORAL STUDENT, I interned at the United States Census Bureau through a National Science Foundation program, which exposed graduate students to working in government agencies to expand their research projects. For my research proposal, I selected the Census Bureau out of the dozen or so federal departments because it has a group called the Center for Disclosure Avoidance Research. The group investigates and develops disclosure control methods, or methods of data privacy and confidentiality to safely release data publicly. I wanted to apply disclosure control methods that used a new privacy definition on various census datasets to provide a proof of concept for these newer methods. Although my proposed methods were well grounded in theory, they had not been rigorously tested on many real-life data.

After being at the Census Bureau for a couple of weeks, the Associate Director for Research and Methodology and Chief Scientist of the Census Bureau called me up to his office. I was terrified. I thought, "What have I done? I haven't been here long enough to do anything!" My fear dissipated some after learning the Associate Director was curious about my research. The United States Census Bureau planned to update their data privacy methodology to use the same data privacy definitions I proposed to explore and test.

DOI: 10.1201/9781003122043-3

As I sat in his office trying not to freak out (he might have given me a heart attack), I could not imagine that years later the United States Census Bureau would decide to implement those new data privacy concepts to protect the 2020 Census—the same concepts that inspired my entire doctoral dissertation.

For this chapter, I will walk through the various ways researchers gain access to data and explain why privacy researchers are still developing new methods.

3.1 WHAT ARE THE PAST AND CURRENT DISCLOSURE CONTROL METHODS?

Figure 3.1: *A Sunday on La Grande Jatte* by Georges Seurat. This image is under Creative Commons Zero, Public Domain Designation from the Art Institute of Chicago.

In Chapter 2, we established a foundation on why institutions, such as the United States Census Bureau, must protect our data. This information allows us to better understand why privacy researchers develop certain statistical disclosure control methods. As promised, there is no math in this book, so to help explain these methods, imagine our sensitive data as the famous painting, *A Sunday on La Grande Jatte* by Georges Seurat (see Figure 3.1).

Although some may be unfamiliar with the name of the painter or the painting, many recognize its' iconic pointillism and figures in a park. Similar to how privacy researchers try to preserve

Figure 3.2: Suppression is a statistical disclosure control method that removes information from the data based on certain criteria. We removal random people to represent suppression.

certain statistical qualities in data, we want people to still recognize the image as *A Sunday on La Grande Jatte* even after we modify it. The alterations or changes we make to the painting will represent a statistical disclosure control method used by several federal statistical agencies [34].

Many of the basic statistical disclosure control methods we encounter are variants of *suppression*, the removal of information based on certain criteria. Most data curators want to remove any personally identifiable information, such as Social Security numbers, before publicly releasing their data for obvious reasons. For our painting, suppression is like removing entire objects or people.

In Figure 3.2, we still see people and understand they are in a park, but not all of them are in the image. Some agencies remove entire groups if they are too easily identifiable by a potential malicious individual. We refer to these people who try to gather information on the confidential data as *data intruders* or *data adversaries*. The Bureau of Labor Statistics, for example, removes healthcare businesses and employee information from a large, sparsely populated county if there are fewer than three healthcare businesses in that county. For the 2010 Census and the American Community Survey Public Use Microdata Samples (i.e.,

(a) Topcoding   (b) Bottomcoding

Figure 3.3: Top- and bottomcoding is a statistical disclosure control method that removes information above a selected upper bound or below a selected lower bound, respectively. We changed the lighter colors to white to represent topcoding and changed the darker colors to black to represent bottomcoding.

detailed population and housing data), the United States Census Bureau has a series of suppression rules. These rules require at least 100,000 records in an area, depending on how detailed the reported values can be, whether the data are repeat over time, and if there are other known public data that contain identical information.

Similar to suppression, we might want to remove extreme outliers in data because they can be easily isolated if they deviate too much from the majority of the data. These outlying records can be unique enough to be identified by data intruders, such as using age where there are fewer people older than 85. Instead of removing them entirely, we can protect these records by *top-* and *bottomcoding*. The Internal Revenue Service topcodes by reporting the age to be "85 or greater" for their public tax data. In other words, topcoding and bottomcoding restrict the values we report to be only so small or so large.

For our painting, we can think of top- or bottomcoding as limiting the color range. Figure 3.3a exemplifies topcoding by setting lighter colors to white at a certain point, whereas Figure 3.3b sets the darker colors to black. In these figures, we no longer see the full spectrum of blues and greens, but we still see the general landscape and most people.

Likely, you, I, and many others would not want our income reported to the cent or single-dollar amount. Neither do data stewards or privacy researchers, because income data tend to be precise. This precision makes the continuous data (i.e., data that can take any value within an interval or range like income data) prone to

Figure 3.4: Rounding is a statistical disclosure control method, in which values are shortened based on certain criteria. We softened or blurred lines around objects in the painting to represent rounding.

record linkage attacks. A data adversary could easily link your data to another database, such employment or housing.

To prevent a data adversary from identifying people, we can round the income values to larger groups instead of reporting the specific single-dollar amount. How we round the values varies by institution and what data that institution are trying to protect. One *rounding* scheme uses an additional randomization for rounding up or rounding down. Suppose someone reported a weekly income of $987, then this income has a 70 percent chance of being rounded up to $990 or a 30 percent chance of being rounded down to $980. Some federal agencies will round income based on where the value falls on a set of ranges. For example, an income of $1,000–$49,999 is rounded to the nearest $100, whereas an income of $50,000+ is rounded to the nearest $1,000.

In Figure 3.4, we represent *rounding* by softening and blurring lines around objects. We know there are people in the park, but their outlines are harder to distinguish.

Another way to protect data is by adding or subtracting random values to the sensitive information. Privacy researchers refer to this approach as *adding noise*, *infusing noise*, *sanitizing* results, or *perturbing* the data. I will use these terms interchangeably

Figure 3.5: Noise infusion or sanitization is a statistical disclosure control method of adding or subtracting random values based on a probability distribution. We lightened and darkened random pixels in the painting by a "coin flip" (i.e., the random chance of getting a heads or tails from flipping a coin) to illustrate noise infusion.

throughout the book, because many statistical disclosure control methods use a version of noise infusion.

So, how do we add noise without completely distorting the data? Privacy experts will often add noise from a probability distribution, such as a bell curve, to preserve certain statistical qualities in the data while changing the values just enough to disguise specific details.

Imagine our income data had age information. We find the first eight records have ages 28, 26, 92, 58, 17, 32, 45, and 53, with a mean age of 44.3. We want to ensure the mean age stays the same (or close to it), so we decide to infuse noise to the ages by adding or subtracting values from a bell curve. Our ages now become 24, 27, 100, 52, 17, 33, 49, and 52, with a mean age of 43.9.[1]

In Figure 3.5, we "flipped a coin" (i.e., the random chance of getting a head or tail from flipping a coin) to decide which pixels of the painting would be lightened or darkened. This change represents us infusing noise to the painting.

---

[1] I generated the values using R, a statistical programming language. The exact code is `set.seed(2) round(rnorm(8, 0, 5))`.

Figure 3.6: Categorical thresholding or generalizing values is a statistical disclosure control method that groups values into broader categories. Limiting the color palette to a single shade of red, blue, green, and so on represents categorical thresholding.

Similar to issues with continuous variables, sometimes data curators collect categorical data (i.e., data that take on discrete values) at very small categories or group levels. Again, finer-grained data increase the likelihood of having unique records, which make them more easily identifiable. To combat this problem, we could broaden the categories. This approach is called *categorical thresholding or generalizing*.

Most federal demographic data reports a person's education level as high school, some college or equivalent, bachelors or equivalent, and graduate degree or equivalent, instead of the exact terminal degree. The United States Census Bureau requires that all categorical variables have at least 10,000 people nationwide in each published category. The categories that do not satisfy this requirement must be combined into broader groups. Many other federal agencies follow the "rule of three." If any combination of categories has less than three records, then the privacy researcher must place those records to a larger group.

We demonstrate categorical thresholding in Figure 3.6 by restricting the color palette to one shade of red, blue, green, and so on.

Figure 3.7: Data swapping is a statistical disclosure control method of switching observations with similar variable characteristics. Randomly swapped sections of the painting represent data swapping.

In 1990, the Census Bureau implemented a new disclosure control technique called *data swapping* for the 1990 Census and continued using it until the 2010 Census. The Census Bureau "swapped" the data between households in different locations with similar variable characteristics. They did not swap all observations, targeting only the records with the highest disclosure risks. Because data swapping applied within a specific geographic area, the approach did not affect the population or other characteristic totals. More importantly, data swapping allowed the census data to be released down to the block level, restoring many of the suppressed tables from past censuses that we went over in Chapter 2.

To help imagine the effects of data swapping, we swapped random sections throughout the image illustrated in Figure 3.7. Privacy experts must select the swap rate carefully, because if the rate is too high, they could generate a random, unrecognizable mess of blocks. The method also does not preserve relationships among multiple variables. In our example, data swapping cannot guarantee the two selected sections for swapping are from the same area of the painting or that both are from similar objects (e.g., boats).

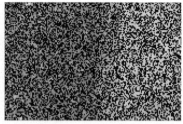

(a) Overall image with 60 percent sampling rate of the pixels.

(b) A zoomed-in image of the sampling 60 percent sampling rate of the pixels.

Figure 3.8: Sampling is a statistical disclosure control method of randomly selecting a subsample of the original confidential data. Randomly sampled pixels at a 60 percent rate represent sampling in the painting.

Data swapping also lacks transparency; the Census Bureau does not report the swap rate because of disclosure risk concerns.

A common statistical disclosure control approach for protecting microdata files that are released to the public, in addition to such methods as top-coding, is to select a subsample of respondents. The U.S. Census Bureau pioneered public use microdata sample files when it released a 1-in-1000 sample of respondents from the 1960 Census. Statistical agencies implement sampling to introduce some form of "plausible deniability" to protect the data. If a data adversary tried to identify a record in our released data, they could not guarantee the match is correct, as the released data are a random subset of the original data.

Suppose we tried to identify someone in the New Mexico employment data as a potential data attacker. The person we want to identify is a Caucasian male, in his mid-thirties, and working as an engineer. We notice a record within the employment data matching this description. However, the person we are trying to identify could plausibly deny the record is his, because the released data are a random sample of the entire population of New Mexico that already has several Caucasian male engineers in their mid-thirties.

Figure 3.8 demonstrates sampling by randomly picking 60 percent of the pixels to keep in the painting. In practice, the sampling rates are usually much lower, like 10 to 20 percent.

In recent decades, *synthetic data*, a technique that generates data with pseudo or "fake" records that are statistically

Figure 3.9: Synthetic data generation is a statistical disclosure control method that aims to create data with pseudo or "fake" records that are statistically representative of the original data. Blurring the background, but not the people, represents generating synthetic data.

representative of the original data, has become one of the most popular statistical disclosure control methods among privacy researchers. Statisticians initially developed synthetic data to address missing data in clinical trial scenarios. Patients would often drop out of studies because they typically lasted for several months or years. The statisticians would fill in the missing values by using a model based on the present or observed data.

Many federal agencies like the idea of synthetic data, because the released data contains no real records, only fake records. Suppose for our example, we want to preserve specific features, such as the people. Figure 3.9 represents a model we select to ensure the people in the park are seen, but we blur the background because maintaining that part of the painting was not a part of our model.

One problem with synthetic data is it requires selecting a model that properly preserves the statistical qualities of the data. This reliance has a few potential drawbacks. Privacy researchers must be careful when selecting and using a model to avoid producing a synthetic dataset that exactly matches the confidential data. If privacy experts incorrectly select a model, then the synthetic data they generate will provide false results for analysts. This also means that developing a model to capture every interesting feature in more complex data without perfectly mimicking the confidential information is extremely difficult.

After learning about data protection approaches, we can more easily understand why marginalized individuals incur a higher privacy cost than others. Most statistical disclosure control methods we covered group, suppress, or randomly change smaller or unique data values. When we alter the data in these ways, we protect the people in the majority more than those who are outliers. The alternative is fully releasing the data unaltered. Both practices are problematic.

Let us revisit the example in Chapter 1, Section 1.4 that looked at how few Hispanic families live in rural Iowa. When data for small subgroups are removed or significantly altered, the people those data represent reap no benefits from later research or public health interventions. If we release data on family composition along with ethnicity for this geographic region, then someone could use our data to identify individual Hispanic families.

3.2 WHAT ARE OTHER WAYS TO ACCESS DATA?

After learning how we could alter *A Sunday on La Grande Jatte*, some might ask, "Why would anyone want to try and access the original data?" After all, despite the alterations we made with our stand-in disclosure methods, we could still recognize the original painting. But, when we examine our altered paintings closely, we see how the statistical disclosure control methods we applied "smooth out" the unique features or outliers from the data. Again, this situation goes back to our data privacy and data usefulness trade-off. Unfortunately, sometimes these features or outliers are essential, such as when studying affordable housing and community development in rural communities.

To gain direct access to confidential data, some institutions establish data use agreements and other legal protections for entities or people outside the original institution that collected the data. Some federal agencies agree to fully release data after a certain amount of time to improve data access. In 1952, for example, the United States Census Bureau and the National Archives entered into an agreement that later became law, known as Title 44, which states that census records containing information about individuals could be released for public use after 72 years. This law is great for balancing security with usefulness, but when a national crisis is happening, such as a global pandemic, we cannot wait 72 years for data access.

Researchers can also access confidential data through Federal Statistical Research Data Centers. These secure facilities provide access to restricted microlevel data (i.e., data at the individual or record level) and are typically located at academic research institutions. As the Census Bureau became more aggressive in their disclosure methods, researchers demanded better access to confidential data.

In response, the Census Bureau established the Center for Economic Studies in 1982 to provide restricted access to economic microlevel data. Researchers had to submit proposals that specified what their economic research was, which economic microlevel data they needed, and how they would publish their results. If the Center for Economic Studies approved of the proposal, researchers became eligible for Special Sworn Status; the ability to work on confidential data, but bound by the same laws as a Census Bureau employee.

In other words, people who obtain Special Sworn Status undergo a thorough background check and, if approved, are "sworn for life" to ensure the privacy of the data they access or face harsh penalties. Other federal agencies, such as the Internal Revenue Service and Bureau of Labor Statistics, also have their own version of Special Sworn Status to allow researchers access to their data.

However, regularly traveling to the Center for Economic Studies in Washington, DC, is out of reach for many researchers. The Census Bureau then established regional Federal Statistical Research Data Centers in several locations, such as Boston, Kansas City, and many California universities. As of this book's publication, there are a total of 30 Federal Statistical Research Data Centers[2] across the United States. Researchers frequently use these centers to access data products from several federal agencies.

---

[2]There will be a new Federal Research Data Center at University of Missouri–St. Louis and will be operational as soon as the 2022–2023 academic year.

3.3 WHY ARE NEW DISCLOSURE CONTROL METHODS STILL BEING DEVELOPED?

We have covered many different methods and processes to release and access confidential data, so your next question may be, "Why are we still developing new statistical disclosure control methods?" If you remember from Chapter 2, the significant evolution of our technological landscape has only made the already difficult problem of balancing data privacy and data utility more challenging.

For example, concerned about the computational advancement, the United States Census Bureau conducted an "attack" on the 2010 Census. Although the Census Bureau used several the statistical disclosure control methods we learned in this chapter on the 2010 Census, they discovered they could re-identify one-in-six of the United States population using publicly available data, such as Facebook [4]. And in 2019, computer scientists from the Imperial College London and Université Catholique de Louvain estimated that they correctly re-identified 99.98 percent of Americans with 15 attributes that included zip code, date of birth, gender, and number of children from anonymized health data [32, 48].

Given this shift in technology and how it impacts data privacy, some might wonder, "Why not allow direct access to the data, such as through the Federal Statistical Research Data Centers and not release public data at all?" Obtaining Special Sworn Status or another equivalent federal background checks typically takes several months or years to complete, not to mention the financial costs of these background checks. And even with the Federal Statistical Research Data Centers located across the United States, these centers favor the top well-connected, higher education institutions and are often too far away for most researchers and data practitioners. For instance, the Rocky Mountain Federal Research Statistical Data Center in Boulder, Colorado is the closest center to me, but I would need to drive more than 400 miles via the interstate. The distribution of these Federal Statistical Research Data Centers calls into question fair and equitable data access for smaller organizations.

In addition to undergoing a background check or clearance process, some data have other restrictions. Federal agencies will often restrict data access to United States citizens or limit the number of nonstaff people who may access the data. Having a public file available avoids some of these data restrictions and reduces the barriers for smaller institutions and local grassroot organizations with less resources.

So what are the new methods being created *now* to improve our access to data? The data privacy community has shifted their focus toward developing and evaluating approaches that satisfy a new privacy definition called differential privacy. I will define differential privacy in the next chapter, but know this definition was the reason why I ended up in the Associate Director's office. The United States Census Bureau, Google, Microsoft, Apple, LinkedIn, and Mozilla are only a few of the many entities developing their own differentially private methods, or algorithms that satisfy differential privacy. Some of these approaches create synthetic data that satisfy differential privacy definitions. Other approaches use an online interface to require individuals to submit their statistical analyses and receive altered results that should be still *close* to the confidential results.

You might wonder, "Why are so many large corporations and federal agencies using differential privacy?" In short, differential privacy makes no assumptions on how a data intruder might attack the confidential data. In contrast, other statistical disclosure control methods must make several assumptions on how a data intruder might attack the public data to quantify the disclosure risk. Differentially private methods, on the other hand, must consider all possible versions of the data that could possibly exist. This means differentially private methods protect the confidential data against data intruders, who arm themselves with any currently unknown or known future information.

This lack of assumptions revolutionized how data curators and privacy researchers protect confidential data. However, privacy experts have mostly focused on theoretical differential privacy work. What few differential privacy practical applications exist are mostly targeting highly complex data systems instead of tackling simpler problems first. These larger data studies require extensive resources and have subsequently faced substantial critiques for their shortcomings.

One example is the 2020 Census. As we have learned, the decennial census is a massive undertaking with significant restrictions on how data must be produced. The data are also used by a large number of public policymakers, practitioners, and researchers. In 2018, the United States Census Bureau decided to shift from traditional statistical disclosure control methods used on the 2010 Census to differentially private methods for the 2020 Census. This quick and pivotal change provided little time for people who use census data to adjust their analyses to account for the differential privacy definition. In response, they have mostly pushed back against the Census Bureau's adoption of differential privacy.

Another source of resistance is the fear they will lose critical components of the data. This fear can be stemmed by two efforts. First, we return to the balance between data privacy and data utility. Stronger privacy protection will naturally result in less information. In 2019, the Census Bureau applied their differentially private method on the 2010 Census as a demonstration dataset for data users to test. Researchers and data practitioners discovered that the demonstration data significantly altered some small population areas compared with the original 2010 Census data, raising concerns for tribal lands and rural communities. Second, most practitioners and public policymakers do not have experience working with differentially private data. In other words, some of their fear around differentially private methods is simply from a lack of experience.

That last piece touches on the additional motivation for writing this book: the demand for more educational materials to better understand data privacy methods. In the next chapter, I will cover in more detail how privacy researchers define privacy, how these privacy definitions differ, how the data privacy community evaluates the data quality, and how they try to balance the two opposing sides.

Expanding access to confidential data is both an art and a science. Privacy experts must carefully consider how to ensure researchers and data practitioners can still draw useful conclusions from data while protecting the privacy of those who may be statistical outliers. That is, we want to retain the image of people enjoying the park but not reveal the boats hidden in the background.

CHAPTER 4

How Do Data Privacy Methods Avoid Invalidating Results?

B EING A PRIVACY RESEARCHER, I am often asked, "How much data privacy protection is enough?" or "How do you ensure that the altered data will still produce valid results?" Most people expect a one-size-fits-all approach can be applied to all data privacy problems. However, my answer is: "It depends."

Sufficient privacy protection could depend on the legal requirements of data stewards, who would shoulder the responsibility for leaked private information. In Chapter 2, we learned how the United States Census Bureau is bound by Title 13. It could also depend on the type of analysis data practitioners plan to implement, such as the Federal Emergency Management Agency (more commonly known as FEMA) needing accurate location data to develop emergency management scenarios. Or, it could depend on how we believe a data intruder will attack confidential data, such as how *New York Times* reporters used cell phone location data to identify Ben Broili, who worked at Microsoft before becoming the manager for Amazon Prime Air.

These are only a few of many factors that shape decisions around applying disclosure control methods. Additionally, those protecting the data must also account for user quality considerations, such as how much, and to what extent, data should be changed. Now we will walk through how the data privacy community navigates the competing needs of data privacy and data utility.

4.1 HOW IS DATA PRIVACY DEFINED?

To protect confidential data, we need to define what we mean by "privacy loss" and how we measure it. As mentioned in the previous chapter, past statistical disclosure control methods define risk differently from differentially private methods, a new privacy loss definition being used by big tech companies. To understand this difference, I will describe how each approach defines disclosure risk and how they diverge from one another. We will start with what I call traditional statistical disclosure control techniques.

Within the traditional statistical disclosure control setting, privacy researchers categorize disclosure risk into three major types: identity, attribute, and inferential [59]. Out of the three, *identity disclosure* is what most people are familiar with. This type of disclosure risk occurs when a data adversary associates a record from an external dataset with another record in the released data. The reidentification could lead to the data intruder learning more information about all the variables in the released dataset with that identified public record.

If you remember from Chapter 1, Section 1.1, Dr. Latanya Sweeney conducted a record linkage attack by linking former Governor Weld's voter registration records to the Massachusetts Group Insurance Commission data. This is an example of identity disclosure. As you might imagine, identity disclosure becomes easier when more external datasets are available.

Another type of attack is when an intruder identifies individual records in data containing outliers or the intruder has specific knowledge of certain key variable values. This situation becomes especially common for continuous variables, such as income. In other words, the data adversary isolates unique records in the data or uses personal knowledge to identify a record in the data, rather than leveraging an external dataset.

Suppose a data curator released income values down to the dollar or cent. A data intruder who knew someone's specific income could easily pinpoint that person in the data. Once the adversary identifies the individual, the adversary can then extrapolate additional information about that person based on the other data variables.

For *attribute disclosure*, a data intruder determines more accurate or new information about a record or group of records based on the structure or features of the released data. Specifically, the data intruder can associate sensitive data characteristics to a particular record or group of records without identifying any exact records.

For instance, if curious but benevolent people of data privacy obtained location data of people's movements before and during a COVID-19 lockdown, they would know which individuals are essential workers and what types of jobs they likely have, based on where they traveled. They could then target any person who follows these habits as an essential worker, even after a lockdown is lifted.

Now, let us build on the data example. Imagine we isolate individuals in the data to people who visit a cardiologist. After analyzing this group of people, we notice most of the records are men older than 65. A data adversary could sell this information to an insurance company. That company could then decide to raise insurance premiums for males older than 65 because they are more likely to have a heart condition. We classify this type of disclosure risk as *inferential disclosure*; when the data intruder infers information about a record with high confidence based on the statistical properties of the released data.

Inferential disclosure has become more important in assessing and protecting confidential data with modern technological advances. More federal agencies disseminate their data and other statistical information through web-based interfaces that allow data practitioners more ease and flexibility to customize statistical outputs. However, this flexibility comes at the cost: less control over what information could be gleaned from the data, causing an increase in inferential disclosure risks [54].

Now that we have defined disclosure risk for traditional methods, we have to measure these risks in confidential and public data. We will start with the most basic metric. First, we would not want too many records with unique combinations of variable values when releasing the data publicly, because data intruders could easily identify them through record linkage attacks or other methods. We can use the simple disclosure measure of "How many unique records are present in the data sample or in the data population?" If we have a high number of unique records, then we have a higher probability that an intruder can determine who or what a particular record represents.

This risk can also be expanded to small groups of two or three records with the same characteristics because they can be as easily identifiable as a single record. Suppose you are trying to hide in a group of a hundred people versus a group of three. This situation is similar to the example in Chapter 1, Section 1.3 when I described myself hiding in Washington, DC, versus in Salmon, Idaho. The privacy risk associated with hiding someone in small groups motivated the development of k-anonymity, a popular disclosure metric among privacy researchers. The k-anonymity metric requires at least one, two, or k many records (so any positive integer value) in any observed combination of identifying variables, such as demographic information in the data.

Another common disclosure metric is estimating the probability of linking a record from publicly released data to external data. The United States Census Bureau, for instance, will release less information on certain variables in the decennial census, given what information they plan to release in the many other data products they produce, such as the American Community Survey. Basically, if these metrics report high probability of identifying a record in the potential public data, then privacy experts might advise to apply some of the traditional statistical disclosure control approaches we learned in Chapter 3 to reduce the risk.

Overall, the general data privacy community views these definitions and other privacy loss measures within the traditional statistical disclosure control framework as more intuitive. In the case of k-anonymity, a large k means we should group that many similar records together to make it more difficult for data adversaries to pinpoint an individual record.

Unfortunately, these intuitive definitions can be somewhat ad hoc. They rely on an accurate understanding of how a data intruder will attempt to extract sensitive information and the external datasets the intruder might use, which includes all possible future data releases. If the privacy researcher incorrectly assumes the data intruder's behavior, privacy protection can be significantly weakened.

Given this problem for traditional statistical disclosure control privacy definitions, our next question becomes, "How does differential privacy define and measure privacy loss?" Essentially, differential privacy throws the traditional statistical disclosure control privacy definition out the window. Differential privacy starts with a "clean slate" and provides very strong privacy protection for confidential data by not making the same assumptions traditional statistical disclosure control methods do.

More specifically, differentially private algorithms do not require an accurate prediction of how the data adversary will attack or what external knowledge that person might use to disclose more sensitive information. Rather, differential privacy sacrifices having an intuitive definition of privacy in exchange for this higher privacy protection, which has challenged and is challenging the data privacy community.

Despite the unintuitive definition, I will attempt to explain how differential privacy defines and quantifies privacy loss at a high level. In general, there are two key points to remember about differential privacy. First, differential privacy is a mathematical definition or condition a method must follow to be *differentially private*. In other words, differential privacy is a statement about the method, not a method itself. This is why I refer to algorithms as "differentially private algorithms" or "algorithms that satisfy differential privacy."

Second, differential privacy uses a conceptual privacy loss "budget," usually denoted mathematically as ϵ,[1] written as epsilon, to help explain the definition in nontechnical terms. The privacy loss budget allows the data steward or others who want to apply a differentially private method to conceptualize how much confidential information is exposed during access.

Basically, if we are data stewards and decide to "spend" more of the privacy loss budget, or a larger value of ϵ, then data users should gain more accurate information during their data analyses. This greater accuracy also means less privacy is guaranteed because more information is being "leaked." Conversely, we could spend a smaller amount of the privacy loss budget, resulting in less accurate information but more privacy protection. As an example, a privacy loss budget of ten means we are releasing more accurate information about the data than if we spent a privacy loss budget of one. Think of ϵ as a knob we can adjust, depending on how much information we want to learn about the data.

In addition to the value we set for ϵ, *sensitivity* of the statistical analysis determines how much noise a differentially private method must add to protect the data. This sensitivity is not in terms of personal or private information, but how robust or resistant the

---

[1] I swear this is the only "math" you will see in the book. In my defense, ϵ is a Greek letter.

information is to outliers' influence. Differential privacy measures this sensitivity by how much the answer to question changes. This change is based on the absence or presence of the most extreme or statistically identifiable possible person that *could exist* but might not be observed in the data.

The reasoning behind this measure accounts for all possible unknowns; because we do not know how the data intruder will attack or what external information they could have, we should protect any *possible version of the data that could exist*. If the answer to a research query is too sensitive to outliers or drastic changes, we should add more noise to protect the data.

Let us walk through an example to illustrate this sensitivity. Imagine the data we want to protect contain demographic and financial information. The question we want answered is, "What is the median income?" According to differential privacy, we must consider the addition or removal of the most extreme possible record that could exist or be in any available data with demographic and financial information. For our example, that person is Jeff Bezos.[2] If Bezos is absent or present in the data, the median income should not change too much. This means we can provide a more accurate answer by adding less noise to the median income result, because the median income question is less sensitive to drastic changes or data outliers, like Bezos. However, if the question is, "What is the maximum income?" Unlike the previous question, the answer would drastically change if Bezos is absent or present. To protect the extreme case (Bezos), a differentially private algorithm must provide a less accurate answer by adding more noise.

Again, differential privacy protects data by assuming any known possible record and that a data intruder could possess any currently known or unknown future information. This last point deviates significantly from the traditional statistical disclosure control definitions, where the traditional definitions often assume the intruder only has information that exists and is currently available. When developing differentially private algorithms, privacy researchers must determine all possible records that could exist within the data, referred to as the universe of possible datasets, to guarantee differential privacy's high standard of privacy protection.

---

[2]According to Forbes, Jeff Bezos is the richest person in the world by net worth for 2020.

Yet, figuring out all the possible records that could be in the data can be difficult if data are not well defined, such as income's unobvious upper limit. Additionally, many privacy experts struggle to grasp the concept of "the entire universe of possible datasets." This situation points to the main drawback for implementing differentially private approaches: it is difficult to understand.

4.2 WHAT IS AN ACCEPTABLE PRIVACY-LOSS LIMIT?

With a basic idea of how we define privacy, we need to decide how to limit or cap the data's disclosure risk. Traditional statistical disclosure control methods usually rely on subject matter experts and public policymakers to give context for the confidential data to help set the limit.

For example, the United States Bureau of Labor Statistics considers employee wages in rural counties as confidential records in employment data. To assess privacy risks, privacy researchers and subject matter experts assume and predict the potential data adversary's background knowledge and behaviors. Again, leveraging expert and public policy knowledge to determine privacy loss is intuitive, but we return to the main criticism for privacy researchers using this type of approach for evaluating disclosure risk. It is too ad hoc.

What about differential privacy? While differential privacy tackles the ad hoc nature of traditional data privacy methods, this definition has a different problem. How do we choose an appropriate value of ϵ for confidential data? Currently, the data privacy community is *still debating* how to pick an appropriate value for ϵ. All we know is, the value must be between zero and infinity. At zero, we either do not release the confidential data or release random values with no relation to those data. At infinity, we release the confidential data without alteration.

When trying to understand the relationship between ϵ and acceptable privacy loss, early differential privacy research suggested that ϵ should be less than or equal to one, and anything greater than two would release too much information. However, most of this research has been theoretical or made assumptions about the data that are not realistic in the real world, such as the ability to categorize all values. Practical applications of differentially private algorithms have shown higher values of ϵ, such as eight or ten. Yet, given the few practical applications, privacy researchers are left asking, "Is eight or ten actually large? Should we expect these values—or even larger ones—when applying differentially private algorithms on more complex, real-world data?"

Meanwhile, those outside the field are wondering, "What does an ϵ value of one, two, eight, or ten even mean?" Without additional context for those who are not experts in the field, these values have no meaning. This situation highlights one of the biggest challenges that *I believe that the field still needs to address*: creating more educational materials for audiences of different backgrounds to understand differential privacy and privacy-loss budgets.

In other words, how do privacy experts describe to public policymakers what it means to use an ϵ value of one versus ten for certain confidential data? While privacy researchers should be contributing to the conversation and decision-making, they must overcome the lack of communication tools and other resources to explain these concepts to non-experts. Additionally, they need to refine the communication concepts that do exist, because many still fall short in actually helping non-experts understand.

For example, privacy researchers often use ϵ as a set "budget" to help people think about managing privacy loss, similar to how people might manage finances. This means there is a "spending limit." But who decides the budget cap for the data? Again, this question becomes a public policy decision for stakeholders, subject matter experts, and data curators. They must determine how they will strike the balance between the need for accurate statistics for various public policy applications and ensuring the data record's privacy, all while converting that risk to an ϵ of one versus ten—and knowing what that means.

An additional challenge for setting a privacy loss budget within a differentially private framework is applying it in practice. Imagine the likely scenario of data users across the United States wanting to access employment data to track job losses during the COVID-19 pandemic. Because job losses have persisted throughout most of 2020 and into 2021, data practitioners will need to repeatedly query or analyze the data. If the system that allows access to the data are differentially private, then the data user must "spend" some amount of the privacy loss budget every time they want information from that data. This information could be an answer to a question ("What is the number of job losses in a region?") or a publicly released altered dataset (e.g., synthetic data). If the United States Bureau of Labor Statistics does not restrict or set a budget cap, then the privacy loss for those employment data will continue to accumulate with each piece of information being released. In this scenario, the privacy budget will "approach" infinity—equivalent to releasing the confidential employment data unaltered.

What if the Bureau of Labor Statistics placed a budget cap? If there is a cap, then the Bureau of Labor Statistics would have to cease all access once that budget cap is reached. This situation is why federal agencies and other entities prefer to generate one synthetic dataset that satisfies differential privacy or other privacy loss measures. By releasing a single differentially private synthetic dataset at a fixed privacy loss budget, data curators avoid "overspending."

However, if public policymakers and data practitioners know the confidential data still exist, they would probably demand continued access to the data in addition to the differentially private synthetic data. This scenario is already being seen with the United States Census Bureau's tentative release of the 2020 Decennial Census. The public data file will be a differentially private synthetic data, but other government officials demand access to the confidential data for certain federal statistics.

As we continue to think about how an entity would enforce an overall privacy loss budget, several more social questions arise. Who should have access to the data? Should there be a limit to the number of data users from the same institution? If selected, how much of the privacy loss budget should be allocated to that person or entity? Should there be a fee to limit access to more serious applicants? How do you discourage people trying to "game the system" once they have access?

These questions bring up potential data access equity issues. In an ideal world, data practitioners from grassroot organizations would have equal access as others from "elite" institutions. In reality, we risk exacerbating data access inequality. In Chapter 2, we discussed how the United States Census Bureau created Federal Statistical Research Data Centers to enable better data access, but these centers are an imperfect equity solution because they tend to be located closer to larger institutions that can support the centers. In Chapter 7, I discuss how the data privacy community is working to answer these outstanding issues

4.3 HOW IS DATA QUALITY ENSURED?

After thoroughly covering the data privacy definitions, we need to assess the data's usefulness, or utility. Similar to questions on defining and measuring disclosure risk, the data privacy community determines data quality by asking, "Who are the data practitioners, and what will they use the data for?" In other words, privacy researchers usually establish data quality benchmarks or metrics based on how people, research institutions, and government agencies will use the data.

However, determining which specific metrics to implement is an entire research field on its own. For ease, I will roughly group the metrics as summary statistics, outcome specific analyses, and global utility measures. But, know there are many types and I am only covering a few.

We will use the United States Internal Revenue Service synthetic tax data on nonfilers.[3]

Many privacy researchers first examine the summary statistics of released data as a quick and easy starting point for assessing utility. Some common summary statistics measure how well the released data preserved the counts, means, and correlations for each variable or combination of variables.

For instance, public policy and privacy researchers developed a synthetic data method to create public tax data. To evaluate utility, they constructed several figures, like Figure 4.1, that show how the original data tax variable averages compared against the synthetic data tax variable averages. The researchers also examined age and gender distribution by comparing the original and synthetic data total numbers or percentages of the variable combinations (see Table 4.1) [6]. In both the figure and table, the original and synthetic data values are close to one another. This indicates the synthetic data captured those utility metrics well.

What if data curators and privacy researchers know who the data practitioners are? They may ask the data practitioners and other data users what specific analyses they typically implement as another measure for data quality. The data privacy community refers to this type of utility measure as an outcome-specific metric.

For our tax data, many public policy research institutions, such as the Urban Institute, implement microsimulation models to determine how new tax policy plans will affect Americans. These models first estimate a baseline from current economic conditions in the United States and then calculate a counterfactual or an alternative estimation based on the proposed policy program change.

---

[3]Nonfilers are individuals who did not file a federal tax return, had no obligation to file, were not claimed as a dependent, but had income reported to the Internal Revenue Service on at least one information return for a given tax year [9].

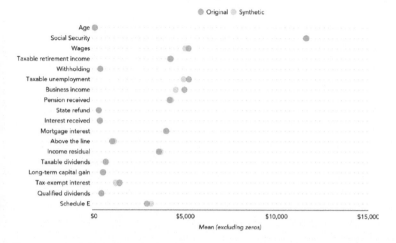

Figure 4.1: The summary statistics means are from the original and synthetic tax data from [6]. We included age in this plot for simplicity, because it is the only continuous variable in the data not measured in dollar amounts.

Table 4.1: Counts of age and gender of the original and synthetic tax data from [6].

AGE	GENDER	ORIGINAL DATA	SYNTHETIC DATA
1–17	Female	412	415
1–17	Male	443	397
18–24	Female	1,030	1,047
18–24	Male	1,268	1,315
25–34	Female	1,090	1,069
25–34	Male	1,717	1,658
35–54	Female	2,494	2,442
35–54	Male	3,587	3,643
55–64	Female	1,892	1,903
55–64	Male	1,994	1,939
65+	Female	6,642	6,529
65+	Male	4,073	4,134

The difference between the baseline and the counterfactual estimates reveals the impact of the proposed public policy program. As an example, the results of microsimulation models are often seen when United States presidential candidates propose new tax policy plans, such as Medicare for All [44, 45].

If we apply these microsimulation models to both the original data and the synthetic data, we can then verify whether the synthetic data will produce similar outcomes compared with the original data. For our nonfiler public tax data, public policy researchers implemented tax microsimulation models that calculated the estimated adjusted gross income, personal exemptions, deductions, regular income tax, and tax on long-term capital gains and dividends on the confidential and synthetic data. Figure 4.2 illustrates that the original and synthetic data produce similar outcomes for adjusted gross income.

Using global utility metrics or discriminant-based algorithms is another way to evaluate data quality. These newer and more complex utility metrics attempt to measure the released data's similarity or proximity to the confidential data. At a high level, these approaches first combine the original data with the publicly released data and mark each record as being from the original data or the altered data. Typically, the altered data are synthetic data.

Next, the privacy researcher must decide what classification model (i.e., a model that predicts whether a record belongs to a specific group or category) to discern if a record is from the confidential or the public dataset. If the classification model struggles to assign a record to either the confidential or public data, privacy researchers then assume that the two datasets are similar. More specifically, each record receives a probability of being classified as being from the confidential data or the public data. A probability is close to 50 percent means the classification model cannot predict any better than a coin flip.

Finally, depending on the global utility measure, the method distills those probabilities into a single value or multiple values to convey how similar the released data are to the original data. The "accuracy" of this type of algorithm depends on what classification model privacy experts use, because each classification model will measure different characteristics of the data. Privacy researchers need to conduct more scientific studies to fully understand these differences.

When measuring the quality of altered data, we must also keep in mind how difficult it is for data stewards and privacy experts

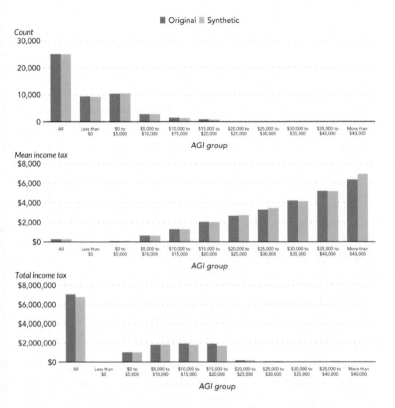

Figure 4.2: The results for calculating the different adjusted gross income groups for count, mean income tax, and total income tax based on the original and synthetic tax data from [6]. AGI stands for adjusted gross income.

to predict all the analyses data practitioners and other researchers will want to conduct on the data. Basically, it is impossible to ensure the released data will provide valid results for *all analyses*. This issue is why data stewards and privacy experts will apply and evaluate a suite of utility metrics to gain a more informative evaluation on the data quality.

4.4 WHY IS BALANCING DATA PRIVACY AND UTIL-ITY *STILL* HARD?

We finally return to the original question posed at the start of this chapter and "It depends." Although we covered the various disclosure risks and utility metrics, we have not discussed how to balance these two competing measures. As a first step toward tackling this difficult problem, we need to ask ourselves, "What is the motivation for developing a data privacy method, who will benefit from the data access, and have we sought input from the data users as you answer these questions?"

While some may believe the first two questions are the same, the latter slightly shifts the focus to what privacy researchers *think* data users want access to versus *why* the data users want access. For the last question, by seeking input from the end user[4] throughout the entire method development process, we avoid creating public data that are unusable for either the intended purpose or broader applications.

To better demonstrate the importance of these questions, suppose we want to publicly publish data on employees and business establishments[5] in the United States. Without a particular end user in mind, we decide to focus on increasing access to more detailed information on business establishments. Our motivation stems from knowing that the past releases of the data only provided 2-digit, instead of 4-digit, North American Industry Classification System codes. United States federal statistical agencies use the North American Industry Classification System, or NAICS codes, for the collection, analysis, and publication of economic data.

For example, 31-33 is the 2-digit code for manufacturing. This code has dozens of 4-digit codes for specific types of manufacturing, ranging from animal food manufacturing to motor vehicle manufacturing. I will refer to these codes as the industry classification codes for simplicity.

Given the drastic difference in granular information from 2- to 4-digit industry classification codes, we decide to try and release the more detailed 4-digit industry classification codes for our data. As we develop our statistical disclosure control method, we

---

[4]End user is a common term in product development that refers to the people who will ultimately use the product or data.

[5]As defined by the Bureau of Labor Statistics, establishments are a single economic unit, such as a mine, a farm, a factory, or a store. Establishments are typically at one physical location and are different from a firm, or a company, which is a business and may consist of one or more establishments.

encounter the privacy and utility trade-off points for our data. We have to suppress some of the data in smaller geographies to keep the 4-digit industry classification code information while reducing disclosure risks. We determine that this trade-off is worth it, because our goal is to expand the industry classification codes.

As we finalize our altered data, we continue to not consult other data users, because we believe we fully understand their needs. For instance, we discover multiple analyses on our own to test the quality of our altered data. One of these analyses investigates the impacts of broadband adoption and broadband program growth for business establishments and employment. We can conduct the analysis within a specific 4-digit industry classification code at the state and county levels.

After undergoing other checks to ensure minimal disclosure risks, we excitedly publish our altered data, expecting to help many data practitioners. Unfortunately, because we didn't consult data users throughout our project, this expectation backfired. We received complaints from multiple researchers that they wanted the employee and business establishments counts at smaller geographies than county level and did not care if the industry classification code had 4 digits or 2 digits.

Although this scenario is hypothetical, misjudging data users' needs occurs more frequently than you might think and can and should be avoided by asking those three guiding questions stated earlier. We will redo this scenario, but this time, we will properly assess data users' needs. Along the way, I will describe how the data privacy community tries to balance data privacy and data utility when releasing data publicly.

We will use the employee and establishment data as our example again. These are a real data product called the Quarterly Census of Employment and Wages, created by the United States Bureau of Labor Statistics. I will refer to this dataset as *employee and establishment data*, as Quarterly Census of Employment and Wages is a bit of a mouthful.

For some background, the employee and establishment data contain information, including counts on the number of employees and establishments, for more than 95 percent of the United States' jobs. Due to privacy concerns, the Bureau of Labor Statistics only publishes the data at the county level, instead of the individual record level, which leads to suppressing over 60 percent of the data. If counties have sparse populations, the Bureau of Labor Statistics also suppresses employee and/or establishment counts at the county level because there are too few records for safe data handling.

This problem motivated my colleagues and I to propose a pilot use case study on developing differentially private methods for the employee and establishment data. After we completed the pilot study, we conducted a literature review and sent surveys to researchers on how they use employee and establishment data to narrow down future research directions. The broadband question posed earlier is one of the survey responses. Economists from the Economic Research Service group in the United States Department of Agriculture wanted to investigate the impacts of broadband adoption and broadband programs in rural economies at the census tract and block groups. This analysis is one of many potential research applications that indicated smaller geographies were more important than detailed industry classification codes.

With the motivation and end user in mind, we can begin the process of balancing data privacy and utility when publicly releasing data.

As a first step, we need to determine the acceptable disclosure risk threshold, the disclosure risk measures we will use, and the utility metrics we will apply to assess the quality of the altered data. Data privacy researchers generally decide the threshold based on who or what is ultimately responsible for the publicly released data.

For the Bureau of Labor Statistics, the agency must follow the Confidential Information Protection and Statistical Efficiency Act when releasing employee and establishment data. Based on those laws or other requirements, we can then decide if we should pursue developing a traditional statistical disclosure control method or differentially private method. As of the publication of this book, the Bureau of Labor Statistics applies traditional statistical disclosure control methods, such as suppression, on the employee and establishment data, but it is actively exploring other data privacy methods to improve access.

As we continue our employee and establishment example, we will broadly discuss both types of privacy-loss definitions. If we want to apply traditional statistical disclosure control methods, we need to decide how we will measure the disclosure risk. Do we include disclosure risk definitions of identity, attribute, and inferential, or some combination of the three? What disclosure risk metrics will we use to measure our selected disclosure risk definitions? One measure the Bureau of Labor Statistics currently uses is the rule of three, where no combination of information can have fewer than three individuals or establishments.

If we decide to use differential privacy, we must choose an appropriate privacy-loss budget. As discussed earlier in the chapter, there is no general consensus on an appropriate value, a major hurdle in implementing differentially private methods. Some privacy researchers will compare a range of ϵ values against various utility metrics and find the optimal balance between the two to establish a privacy loss budget. This type of approach directly leads to the need to balance and assess the selected privacy loss definition against a suite of data quality metrics. Again, most privacy researchers base these utility metrics on analyses data practitioners typically apply in addition to other statistics.

Next, we need to eliminate variables that are too sensitive to release. This is usually decided by the data stewards, though sometimes privacy researchers help. They may suggest additional variables to remove that are seemingly harmless to keep but can reveal confidential information indirectly, such as our cardiologist visit example. For the employee and establishment data, the Bureau of Labor Statistics automatically removes employee and business establishment names and other demographic information.

Based on the variables we removed, we can decide if all the remaining variables should be included when developing our data privacy methods. Another question we should consider is, "What parts of the data are essential, desired but not necessary, and unnecessary to accomplish our goal?" In other words, knowing what variables or other information in the data are of high, medium, and low priority will help us determine what parts of the data need to be altered and to what extent. For our example, a high priority would be preserving the number of employees and business establishments for smaller geographies, whereas more specific industry classification codes—such as manufacturing versus animal food manufacturing—would be a lower priority.

Finally, we are at the point where we develop and implement statistical disclosure control methods and evaluate the data quality with our selected utility metrics. Essentially, data curators and privacy researchers fine-tune their statistical disclosure control methods by repeatedly evaluating if the altered data are at acceptable levels of disclosure risk and quality. Often, the process of adjusting the statistical disclosure control methods and earlier steps in the process becomes analogous to "holding sand." Shifting or changing one part of the workflow, such as trying to improve the data quality for one variable, can result in the privacy "spilling out" in unexpected ways.

The rule of three, for instance, is one of the reasons why over 60 percent of the employee and establishment data are suppressed. Although the rule is applied after two other disclosure risk measures, the rule of three prevents the majority of the employee and establishment information from being released publicly.

From walking through this example, we can summarize this process or workflow as six basic steps.

1. Determine the threshold of acceptable disclosure risk, disclosure risk measurements, and utility metrics.

2. Remove personally identifiable information and variables that are too sensitive to include in the final released data.

3. Select the variables in the confidential data that need to be altered, which are usually all remaining variables.

4. Develop and apply the statistical disclosure control methods that will reduce the specific disclosure risk measurements and will preserve the chosen utility metrics.

5. Compare the data quality of the released data to the confidential data using the utility metrics.

6. Repeat steps 4 and 5 if the disclosure risks are too high and/or if the data utility results are too low.

4.5 WHAT IS THE DATA PRIVACY FRAMEWORK?

Although the workflow seems simple enough, step 4 is a "loaded step." Even for privacy experts, *correctly* developing and applying statistical disclosure control methods can be quite difficult and grows significantly more complex depending on the data structure itself and data practitioners' needs for the confidential data. These problems frequently occur when privacy researchers do not follow a framework or track how information flows from confidential data to data users. Not following a framework or having some means to track the flow of information can lead to improper data privacy methods or methods that are too messy and obscure for others to understand.

Both outcomes are equally bad. When I was a naive first-year doctoral student, I excitedly started my data privacy literature review for my dissertation with a focus on differential privacy. At this

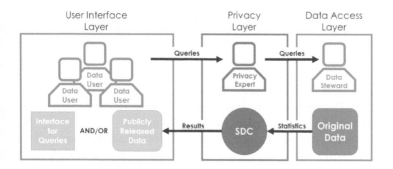

Figure 4.3: Diagram of the statistical disclosure control (SDC) framework.

time, differential privacy had only been around for a few years, so I quickly discovered how limited the written materials and other resources were on differentially privacy. I remember scouring the internet for *anything*; blogs, articles, course notes, and even professors' musings. What materials I did find were a grab bag of quality and clarity.

On the quality side, because I was new to the field, I could not know if a researcher had created an approach incorrectly. By learning from the wrong applications, I got confused and had to be corrected by my doctoral advisor. On the clarity side, when I learned from applications that suffered from a messy framework, I felt like the researcher threw paint on the wall to see what stuck, and I had to interpret the smatterings. Sometimes I understood why the researcher smudged paint in one area, and other times I completely misinterpreted their intent and internalized a wrong data privacy concept.

Since those dark times during my graduate career, data privacy experts have been more consistent in creating differentially private methods. We can now break down how differential privacy algorithms are constructed into a similar framework and workflow as many traditional statistical disclosure control approaches. While the framework that I will describe does not provide exact details on how a privacy researcher develops a data privacy method, it will cover the basic structure and how to tell when a method is done right or wrong.

In Figure 4.3, we see the framework having three basic "layers": user interface, privacy, and data access. Within the user interface layer, data practitioners and users pose the questions, queries, or analyses they want to apply to the confidential data. Privacy experts gather these queries within the privacy layer and work with the data stewards with direct access to the original data. Together, they pinpoint the key statistics in the confidential data users want, determine how to preserve those qualities, and select the statistical disclosure control technique that balances their privacy requirements against the practitioners' utility needs.

Any statistics or information that leave the data access layer must enter the privacy layer to be altered. The modified statistic can then leave the privacy layer to be accessed by data users as part of the publicly released data and/or through an interactive interface, like a web-based dashboard. We can breakdown the transition of information between the data access layer and the privacy layer into three basic steps:

1. Pre-processing step: Determining priorities for which statistics to preserve

2. Privacy step: Applying a statistical disclosure control technique to the desired statistic

3. Post-processing step: Ensuring the results of the statistics are consistent with realistic constraints, such as negative population counts

Although this framework should be easy to follow, privacy researchers who create incorrect methods tend to not track how *all the information* flows through each of the layers. To avoid this problem, remember: statistics that leave the data access layer must immediately enter the privacy layer. From there, the privacy expert must alter the statistic using a statistical disclosure control method before being released to the user interface layer. Oftentimes, improper methods release information from the data access layer directly into the user interface layer, skipping the privacy layer completely. Statistical disclosure control methods that do not follow this overall framework should raise a red flag.

4.6   TO POST-PROCESS OR NOT TO POST-PROCESS?

Another major and highly debated challenge in developing a statistical disclosure control method is how researchers enforce the post-processing step. If they are not careful, privacy researchers might

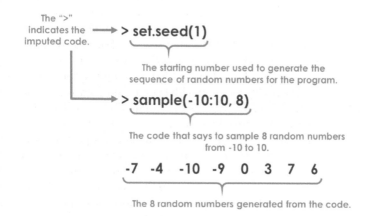

Figure 4.4: Diagram explaining the code I used to generate random numbers from −10 to 10.

worsen inconsistencies they are trying to correct or cause grossly inaccurate results when data practitioners apply their analyses.

Let us go back to the employee and establishment data example. Suppose our goal is to release the number of people working at manufacturing establishments within a county with certain demographic information: age, gender, and race. One possible group would be African American men ages 50 to 55. To protect these data, we decide to apply random noise. As we add the noise, we encounter inconsistency issues. The number of African American men in their early 50s has been changed to a negative count. We cannot release the data as is—negative counts are not physically possible. We are left with the choice of either changing the count to zero or resampling the noise being applied.

Both options have pros and cons. Setting the count to zero is quick and simple but eliminates an entire subgroup that may cost unintended, disproportionate privacy losses for an underrepresented group. Picking this method could also skew the data with *several* zero counts, introducing more biases into the data. If we decide to post-process the data for all United States counties in this manner, a potential, though extreme, outcome could be that certain groups become nonexistent and others are overrepresented.

For the other option, resampling the noise could add high computational time costs. The increased computing time usually occurs when we are trying to sample values that are physically possible

Table 4.2: An example to show how the sum of noisy or altered counts might not sum to the expected total and how some privacy researchers readjust the counts

TRACT	ORIGINAL	ALTERED	READJUSTED
Tract 1	15	20	$100 \times (20/105) \approx 19$
Tract 2	75	70	$100 \times (70/105) \approx 67$
Tract 3	10	15	$100 \times (15/105) \approx 14$
Total	100	105	100

but have a low probability. This situation means we could be re-sampling multiple times until we obtain a value that is physically possible.

For our example, imagine there were only two African American males in their early 50s in the original data for a particular county. The noise we are adding ranges from −10 to 10. If we draw any values from −10 to −3, we will need to resample because we will have a negative count. Figure 4.4 shows code[6] I used to randomly draw eight values between −10 to 10. In the figure, we see that the computer had to randomly sample the values five times until we added a random value of 0 to our true count of 2. Although sampling five times is very quick for most computers, there are other more complex data structures where resampling on even a very powerful and fast computer can take several hours until a *single* appropriate value is drawn.

Continuous variables also have similar consistency issues. What if we wanted to publish statistics of employee wages for the county? Similar to the count example before, we cannot have a negative wage because it is not a real-world possibility. If we decide to alter some of the wages, we need to be aware of the same issues we encountered for the count data and ask similar questions, like, "Do we set the bad wage value to a known bound?"

Another post-processing issue we might encounter is the sum of parts must equal the whole. In this situation, suppose we altered the total establishment counts for three census tracts. If we decide to modify these groups separately, we could not guarantee that the sum would equal the original county total, like we see in the

---

[6]The code is from the statistical programming language called R.

third column of Table 4.2. Some privacy researchers might leave the changed counts as is, because, for this example, the total number of counts for the county would be indirectly protected. Other privacy researchers might suggest readjusting the values based on the county total in the original data, as we see in the last column of Table 4.2, because that value cannot be changed for a specific analysis or other reasons. For instance, the United States Census Bureau cannot alter the population counts for states because an accurate count is required for allocating congressional seats for the House of Representatives.

These post-processing and consistency examples highlight the tip of the iceberg of issues surrounding whether altered data should or should not be post-processed. On one hand, post-processing introduces unintended biases. On the other hand, not post-processing causes confusion for many data practitioners, who would wonder why their data have weird values or violate other rules and laws, such as requiring an accurate population count for congressional seats.

Whether the published data should be post-processed or not be post-processed must be considered carefully, as with all other parts of the process to safely expand data access. There is no "one-size-fits-all" in finding the perfect balance between data privacy and data utility.

CHAPTER 5

What Makes Datasets Difficult for Data Privacy?

BEFORE I BECAME A STATISTICIAN, I wanted to be a physicist. During my time as a physics student, one of the classic jokes I learned is "assume a spherical cow." This phrase originated from the common practice of transforming complex physical problems into more simplified ones. Physics students can then solve these easier problems first before they tackle the original problems later in their education. A common example is, "An airplane is moving from east to west at X miles per hour and is Y feet in the air. If we drop a cow from the airplane, then what is the speed of the cow when it hits the ground?"

When my physics professor first introduced a spherical cow problem, many of my classmates and I laughed at the thought of dropping a spherical cow from an airplane. However, my professor's goal was not to amuse the class. He instead wanted to make calculating the rate of the falling object easier. By assuming the cow was spherical and in a vacuum, we can use the simplified drag equation and Newton's Laws. If we do not, we must use the full drag equation, account for wind resistance, include variations in air density as the cow falls closer to the ground, and several other physical factors. As a new physics student, I would rather solve the former problem than the latter.

While imagining spherical cows may be entertaining, you might wonder why I am talking about them in the first place. After all, this is a book on data privacy and statistics. Not cows and obscure physics problems. Similar to my physics training, I had to start with something simple, like a spherical cow, before considering all the pieces that go into developing a statistical disclosure control method. I learned to apply basic data privacy techniques on simple data, such as data with few variables, to ensure I understood the key concepts before tackling more complex data.

However, just like cows are not spherical, real-life data are not simple. Imagine we collected people's smartphone information. We could learn how often, what form, and at what time people contact others, such as texting more frequently on weekends. Based on these contacts, we could figure out if any of them have similar last names or live close to one another. If the smartphone users installed other apps, we can collect additional information, like what music they listen to or what shows they watch. From their favorite web browser, we could also determine what web pages they visit or if they saved an account on them. Finally, we could track the person's movement throughout the day, capturing their time and location information.

Knowing all that, take some time to think about the following questions:

- What basic steps would you do to protect a smartphone user's data privacy?

- When thinking through the previous question, do you think the smartphone data are harder or easier to protect than some other data we have discussed about so far?

- Why do you think the other data are harder or easier to protect than the smartphone data?

The last question can be difficult to think through, so let us compare the smartphone data to another dataset together. We will use the 2020 Census data for our comparison. Similar to the smartphone data, we must know what data are collected on the 2020 Census. If you resided within the United States on April 1, 2020, the law required you to fill out a form resembling the one in Figure 5.1. The form consists of twelve questions, ranging from how many people live in your household to what is your race and ethnicity. The Census Bureau also collected information on your physical location on April 1.

Start here OR go online at [url removed] to complete your 2020 Census questionnaire.
Use a blue or black pen.

Before you answer Question 1, count the people living in this house, apartment, or mobile home using our guidelines.

- Count all people, including babies, who live and sleep here most of the time.
- If no one lives and sleeps at this address most of the time, go online at [url removed] or call the number on page 8.

The census must also include people without a permanent place to live, so:

- If someone who does not have a permanent place to live is staying here on April 1, 2020, count that person.

The Census Bureau also conducts counts in institutions and other places, so:

- Do not count anyone living away from here, either at college or in the Armed Forces.
- Do not count anyone in a nursing home, jail, prison, detention facility, etc., on April 1, 2020.
- Leave these people off your questionnaire, even if they will return to live here after they leave college, the nursing home, the military, jail, etc. Otherwise, they may be counted twice.

1. How many people were living or staying in this house, apartment, or mobile home on April 1, 2020?

Number of people =

2. Were there any additional people staying here on April 1, 2020 that you did not include in Question 1?
Mark ✗ all that apply.

☐ Children, related or unrelated, such as newborn babies, grandchildren, or foster children
☐ Relatives, such as adult children, cousins, or in-laws
☐ Nonrelatives, such as roommates or live-in babysitters
☐ People staying here temporarily
☐ No additional people

3. Is this house, apartment, or mobile home — Mark ☐ ONE box.

☐ Owned by you or someone in this household with a mortgage or loan? Include home equity loans.
☐ Owned by you or someone in this household free and clear (without a mortgage or loan)?
☐ Rented?
☐ Occupied without payment of rent?

4. What is your telephone number?
We will only contact you if needed for official Census Bureau business.
Telephone Number

Person 1

5. Please provide information for each person living here. If there is someone living here who pays the rent or owns this residence, start by listing him or her as Person 1. If the owner or the person who pays the rent does not live here, start by listing any adult living here as Person 1.

What is Person 1's name? Print name below.
First Name   MI

Last Name(s)

6. What is Person 1's sex? Mark ☐ ONE box.
☐ Male   ☐ Female

7. What is Person 1's age and what is Person 1's date of birth? For babies less than 1 year old, do not write the age in months. Write 0 as the age.
Print numbers in boxes.
Age on April 1, 2020   Month   Day   Year of birth

years

→ NOTE: Please answer BOTH Question 8 about Hispanic origin and Question 9 about race. For this census, Hispanic origins are not races.

8. Is Person 1 of Hispanic, Latino, or Spanish origin?

☐ No, not of Hispanic, Latino, or Spanish origin
☐ Yes, Mexican, Mexican Am., Chicano
☐ Yes, Puerto Rican
☐ Yes, Cuban
☐ Yes, another Hispanic, Latino, or Spanish origin – Print, for example, Salvadoran, Dominican, Colombian, Guatemalan, Spaniard, Ecuadorian, etc.

9. What is Person 1's race?
Mark ✗ one or more boxes AND print origins.

☐ White – Print, for example, German, Irish, English, Italian, Lebanese, Egyptian, etc.

☐ Black or African Am. – Print, for example, African American, Jamaican, Haitian, Nigerian, Ethiopian, Somali, etc.

☐ American Indian or Alaska Native – Print name of enrolled or principal tribe(s), for example, Navajo Nation, Blackfeet Tribe, Mayan, Aztec, Native Village of Barrow Inupiat Traditional Government, Nome Eskimo Community, etc.

☐ Chinese   ☐ Vietnamese   ☐ Native Hawaiian
☐ Filipino   ☐ Korean   ☐ Samoan
☐ Asian Indian   ☐ Japanese   ☐ Chamorro
☐ Other Asian –   ☐ Other Pacific Islander –
Print, for example,   Print, for example,
Pakistani, Cambodian,   Tongan, Fijian,
Hmong, etc.   Marshallese, etc.

☐ Some other race – Print race or origin.

→ If more people were counted in Question 1 on the front page, continue with Person 2 on the next page.

Figure 5.1: An informational copy of the 2020 Census questionnaire.

Although I described the 2020 Census data as a massive undertaking earlier in the book (and it is), many privacy researchers would consider this data "easier" to tackle than the smartphone data. Why is that? I will first describe the census data and then compare it to smartphone data.

The 2020 Census contains a total of twelve questions. A person's answer is limited to the check boxes or fill in the blank boxes, restricting the possible outcomes. The form does not ask how the individuals within the household are related to anyone else filling out their own form. Privacy researchers will also only need to consider a few constraints, such as the geographic information for the post-processing step. For instance, referring to Figure 2.2, one constraint is ensuring the population counts at each census geographic level should sum to the totals at higher levels. Finally, the United States Census Bureau only collects this data once every ten years, and they do not link any records to other previous census datasets.

In contrast, we realize that smartphone data gathers more detailed personal information every day that can be easily tied to other individuals. In other words, the smartphone data has far more information being collected than the 2020 Census. This information also comes in different forms (e.g., web browser searches), vary in how they are correlated with one another (e.g., social media use), and how they should be constrained based on physical properties (e.g., how quickly someone travels from one location to the next). To complicate the situation further, some constraints to the data could make ensuring privacy more difficult. A particular constraint could introduce additional information indirectly. The path someone takes, for example, from one point to another could be limited by geographic barriers, such as lakes or mountains.

In this chapter, I will use myself and a theoretical Asian American family of four as examples to help explain why some data structures are more challenging to protect than others. The different data structures I cover will not be comprehensive. Rather, I will provide a general picture of the real-world challenges the data privacy community faces. Additionally, I will describe how privacy researchers transform the complex and challenging data into a "spherical cow" to make applying statistical disclosure control methods easier.

5.1 WHY DOES CONTRACT TRACING COST PRIVACY?

We will follow our theoretical Asian American family of four daily routine to help us better understand the data privacy and data utility trade-off for spatial data. Suppose our family installed a contract tracing app on their smartphones. This app collected their demographic information and tracked their movements throughout the day. Similar to many proposed and developed contract tracing apps, the company that created the app gathers and stores all this information to notify other users quickly if anyone is within six feet of someone who tested positive for the coronavirus and has been exposed for more than 15 minutes. The company then uses the extra demographic information to learn if certain groups are more affected by COVID-19 than others.

Although society will greatly benefit from this data by slowing the spread of the coronavirus, we can infer a lot of personal information about this family. As mentioned in Chapter 1, we could easily figure out where the kids go to school, what extracurricular activities they do, where they like to hang out, and when they are likely to be home alone. In the wrong hands, this type of data could empower malicious people, such as scammers, to more easily identify and take advantage of the children or other vulnerable members of the family.

Some might think the demographic information is obviously too much. What if we removed it from the data and only kept the individual movements? A data intruder could still easily identify the Asian American family based on the location data. The intruder, for example, could isolate a few individuals in the data and determine their routine based on where they are located at certain times throughout the day. The data intruder could then assess where that person works or goes to school and whether that person changed their routine. Sound familiar? This is the example of how *The New York Times* reporters discovered a former Microsoft engineer, who changed his routine to interview at Amazon.

Another issue is not being aware of possible negative outcomes that data users might draw from the spatial data. Again, predicting what people will do with the data is difficult. But, the data stewards, privacy experts, and data practitioners must try. They have an ethical obligation and responsibility to ensure the released data will not cause harmful or negative ramifications.

Suppose a travel company has an interactive dashboard to show their past customers "exotic trips" to encourage more people to travel. Due to data privacy concerns, the travel company aggregates their data to display the overall percent increases or decreases to certain locations around the world. The company also includes basic demographic breakdown of their customers to show where certain people like to travel.

Imagine you are accessing the dashboard in early 2020, and you observe an increase of Asian Americans traveling to China from the United States and back. What is your reaction to this information? Some might think about how the coronavirus first spread throughout the United States. In 2020, Chinese New Year was January 25 to 31 and is a popular time for many Chinese Americans to visit family. This time is also when the coronavirus started to hit East Asia hard.

Although I presented the travel data as a theoretical scenario, the United States experienced a significant increase in hate crimes against Asian Americans, such as the Korean woman who was assaulted in New York City and Asian American children being bullied more at schools across the country [33]. What if someone saw that travel data and decided to infer their neighbor, our Asian American family, had likely traveled to East Asia. What is to stop that person from harassing the family?

From these examples, we see how adding location and time to the data, especially with demographic information, can cause more difficulties in protecting people's data privacy. We also learned how we must be careful in how we present that information to avoid unintended harm. Given the difficulties of releasing geospatial data, some might ask, "How is the data privacy community releasing that information? What is their spherical cow equivalent?"

In 2006, the United States Census Bureau launched On-TheMap,[1] an interactive web-based dashboard that provided economic and location information on approximately 8 million census blocks. Data practitioners used OnTheMap for planning transportation infrastructure, outlining emergency services, and developing economic programs for growth. The Census Bureau tested and updated their privacy protections several times until 2008, where they applied a method that satisfied a version of differential privacy. When implementing the method, the privacy

---

[1] You can check out the OnTheMap dashboard at https://onthemap.ces.census.gov/.

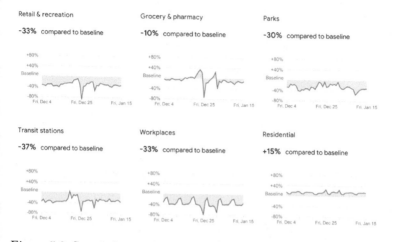

Figure 5.2: Screenshot of the Google COVID-19 Community Mobility Report for Santa Fe County from December 4, 2020 to January 15, 2021. The plots show average movement increase or decrease for each category from the baseline. Google calculated the baseline as the median value for the corresponding day of the week, during the 5-week period January 3 to February 6, 2020.

researchers simplified the complicated data in several ways, such as limiting the possible routes a person would travel and coarsening geographic areas if certain conditions applied. The Census Bureau also enforced certain suppression rules if some areas of the data were deemed too sensitive.

More recently in 2020, Google created the COVID-19 Community Mobility Reports[2] that also implemented a method that used differential privacy. These reports provided movement trends over time at different geographic regions for various categories, such as transit stations and workplaces. They simplified the geospatial problem by releasing the information at larger geography levels (e.g., county) and with no other types of information (e.g., demographic information). Figure 5.2 shows the movement trends in Santa Fe County from December 4, 2020 to January 15, 2021.

---

[2]You can access these reports at https://www.google.com/covid19/mobility/. The site states that, "These reports will be available for a limited time, so long as public health officials find them useful in their work to stop the spread of COVID-19." The website was still up while I wrote this book.

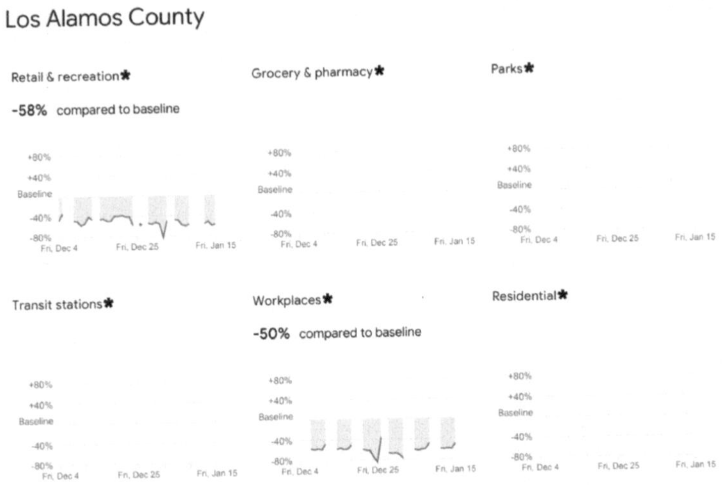

Los Alamos County

Figure 5.3: Screenshot of the Google COVID-19 Community Mobility Report for Los Alamos County from December 4, 2020 to January 15, 2021. The plots show average movement increase or decrease for each category from the baseline. Google calculated the baseline as the median value for the corresponding day of the week, during the 5-week period January 3 to February 6, 2020. The * indicates that "The data doesn't meet quality and privacy thresholds for every day in the chart."

Similar to OnTheMap, the COVID-19 Community Mobility Reports suppressed some of the information if the data does not meet certain quality and privacy standards. Figure 5.3 displays only some data for Retail & Recreation and Residential in Los Alamos County, which borders Santa Fe County. One reason for the suppression is likely the smaller population of 19,369 compared to Santa Fe County's population of 150,358.[3]

OnTheMap and the COVID-19 Community Mobility Reports demonstrate how much progress the data privacy community has made on providing better location data and how much these methods still need to be improved. In late 2020, the National Institute of Standards and Technology Public Safety Communication Research Division launched a crowd-sourcing challenge called the Differential Privacy Temporal Map Challenge to address some of the persisting

---

[3]United States Census Bureau population estimates for 2019.

space and time data problems. The data challenge aimed to encourage competitors with cash prizes to develop new and innovative differentially private methods that applied to emergency planning, spatial data. It helped push the boundaries of the field further to foster more practical, data-driven research and development.

5.2 WHY DOES MEMORY FADE OVER TIME BUT PRIVACY DOES NOT?

In our last section, the examples we covered included a time element since we were tracking when individuals or groups move from one location to another. This situation brings up the question of, "If we isolate the data based on time with no spatial component, then would protecting time information be easier?"

Again, let us explore an example to see what are the possible issues. In Chapter 1, I mentioned that I have an "unhealthy obsession" with participating in triathlons. Since my first year in college, I have raced in several run, cycle, and triathlon events. Imagine a data intruder decided to gather my information that is publicly available on the various race event websites. They would learn:

- **How old I am:** The data intruder could pinpoint my exact age based on when I changed Age Groups from one year to the next. For example, someone being in the 40–44 Age Group one year and then being in the 45-49 Age Group the following year.

- **When I married:** Most women in the United States change their last name to their spouse's. An intruder could learn roughly the year I married by finding when my last name changed from one racing event to another.

- **Where I lived and when:** Many race events list the participants' hometown to show the geographic diversity of people at the race. The data intruder could assess when I moved from Idaho to Indiana and from Indiana to New Mexico.

Based on what we can learn from race event websites, some of you might think these websites should not reveal so much data on a person. However, a single race typically contains limited information. For instance, the full-distance triathlon I completed only lists my name (i.e., first and last name), country (i.e., USA), division (i.e., Gender + Age Group), finishing time (i.e., hours:minutes:seconds), and ranks (i.e., rank within division) on the website. The reason we learned so much information about me is because we linked multiple races together across time. Researchers refer to this type of data as longitudinal data, where we have repeated observations of the same person or record over time.

You likely noticed that we indirectly gained a lot of additional information. For privacy experts, trying to prevent this type of disclosure risk makes protecting time data even harder. One common and unsatisfying statistical disclosure control approach is not linking the data at all. In other words, the data are repeatedly collected with no observations linked across those datasets. As mentioned before, the United States Census Bureau does not publish any identifiers for the census data even though the data are collected every ten years.

This approach has an obvious drawback; the data are no longer longitudinal. We then lose potential insight into the data to make more data-informed decisions. For instance, the United States Federal Government uses the census to reallocate congressional seats. With the data, practitioners can see where the population has increased or decreased at certain geography levels.

Suppose we focus on Idaho. The Census Bureau reported Idaho as the fastest growing state by percent population change from 2017 to 2020 and the second fastest growing state from the 2010 Census to the 2020 Census—right behind Utah. The Census Bureau cited domestic migration as the main cause for the population increase. However, data practitioners will not know the state or demographic breakdown of who is moving or leaving Idaho. Given the importance of congressional seats, state governments want to identify the causes for their population shifts to maintain or gain congressional representatives.

Not correlating datasets across time causes another issue— possible inconsistent or misleading trends. The Bureau of Labor Statistics, for example, publishes the employee and establishment data[4] every three months and applies their statistical disclosure control method independently on each dataset. Imagine that in one quarter the data showed there were seven establishments in a county, the next quarter there were five, and in the third quarter there were seven establishments again. In the three datasets, we saw what the Bureau of Labor Statistics calls the *deaths* and *births* of two establishments. However, without identifiers to link

---

[4]The Quarterly Census of Employment and Wages Data.

the establishments, we do not know if the births and deaths of the establishments were due to seasonal change, such as a tourism, or a statistical disclosure control method that removed the two establishments due to privacy concerns for that particular quarter.

Overall, certain trends should hopefully be preserved as data curators and privacy experts follow the data privacy workflow from Chapter 4, Section 4.4. We know from this same workflow that balancing the need for data privacy and data utility is difficult and even more so at smaller data sub-populations.

In an attempt to give this spherical cow "some legs," privacy researchers will sometimes create larger groups out of the data as a step towards keeping the data linked across time points. The data stewards release the data by grouping people together based on similar characteristics or aggregating the data into larger time spans. Researchers sometimes refer to these larger groups as cohorts. We will use education data as an example to explain both the similar characteristics cohort and larger time span cohort versions.

Imagine the Department of Education in federal and state governments want to know how certain education policies affect students. The data stewards could identify the years a certain policy has been or not been in effect for certain students and group them into policy based cohorts. Data practitioners could then track these cohorts over time to discern specific outcomes, such as improved math and reading scores.

For time span cohorts, suppose the Department of Education also wants to know how the student population progresses through the education system to meet workforce demand. The data stewards could focus on the graduation rates information. They could increase the time spans based on major milestones or physical constraints, such as reporting when a student is in high school and transitioning to trade schools, colleges, and other institutions. With this type of data, users can analyze trends in how people complete their education before starting their first job.

However, these larger cohort sizes are still not ideal. Many researchers still want access to the record or micro-level data for specific statistical analyses. More recently, the United States Census Bureau in collaboration with Cornell University created a system to allow researchers access to two synthetic Census Bureau data products and a mechanism to validate their results on the confidential data without "seeing" the confidential data. The two synthetic datasets are the Synthetic Longitudinal Business Database, a synthetic longitudinal dataset of establishment-level business microdata starting from 1976, and the Survey of Income and Program Participation Synthetic Beta, a synthetic data that integrates person-level micro-data from a household survey with administrative tax and benefit data.

Researchers wanting access to the two synthetic datasets must first submit a request to Cornell University, who hosts the data on their Synthetic Data Server.[5] This server is a restricted-access server that can be logged on from anywhere that has an internet connection, but researchers are prohibited from transferring programs or data to and from the server. Once granted access, researchers can apply their statistical analyses on the synthetic data. During this process, researchers are cautioned not to fully trust the statistical results they observe from the synthetic data without validating them against the confidential data.

Some might wonder, "Isn't the purpose of this synthetic server to prevent people from accessing the confidential data? How would a researcher validate their results?" The researcher avoids seeing the confidential data by having someone else apply their analyses on the confidential data. The researcher submits their analyses that were tested on the synthetic data through the Synthetic Data Server portal, where an approved staff member will apply the analyses on the confidential data. That staff member will then review the confidential data results to ensure it passes the Census Bureau disclosure rules for external projects before sharing the result to the researcher.

Other statistical federal agencies are developing or looking into this two-stage approach for allowing researchers more accurate analyses while still preserving the data privacy. Major shortcomings include the manual and subjective vetting of results that take a minimum of a week and upwards to several weeks depending on several factors, such as computational time. Despite the issues, this type of system will likely be more prominent in future data privacy research, which I will cover in Chapter 7.

One other solution privacy experts propose is to create alternative privacy definitions that are not as strict as the original differential privacy definition. When privacy researchers started developing these new privacy definitions, they started to refer to the "original" definition as pure differential privacy and the alternatives as

---

[5]You can request access to the server at https://www2.vrdc.cornell.edu/news/synthetic-data-server/.

relaxed differential privacy definitions. While there are several different relaxed differential privacy definitions, I will focus on one definition since it will come up in multiple examples.

A popular alternative or relaxed differential privacy definition is called local differential privacy, which has the same spirit as pure differential privacy but is conceptually different. Local differential privacy focuses on how individual data are collected and aggregated for analysis. A local differentially private method will sanitize or add noise to each individual's data before sending it to the data curator. This means local differentially private methods trust no one, including the data curator.

Using this definition, Google created a local differentially private method called RAPPOR (Randomized Aggregatable Privacy-Preserving Ordinal Response), an end user client software on Chrome browser data for crowd-sourcing statistics. When Google initially released RAPPOR, the examples in the original research paper were idealized and they did not fully track the privacy loss budget. RAPPOR was one of the first major methods applied to longitudinal data within a differentially private framework and is used as a starting point to improve similar methods for future research.

As the privacy experts continue to develop better ways to access longitudinal data, a major challenge is from legal aspects and not from the methodology side. In other words, privacy laws may change over the course of the data, requiring different rules on the data collection and dissemination. With little research in how to handle the intersection of data privacy law and data privacy methodology, the data privacy community should and must focus on how to integrate these two sides more seamlessly or at least create a set of guidelines for best practices. More on this in the next chapter.

5.3 WHY ARE PERSONAL RELATIONSHIPS COMPLICATED?

Social network analysis and research has also significantly grown in popularity; driven with the emergence of social media, such as Instagram and Twitter. Social network data applications have included tracking Twitter accounts to monitor mental health, evaluating and predicting knowledge flow within an insurance company, determining the affects of social network sites on the learning and teaching of K-12 students, and analyzing social distancing strategies for flattening the COVID-19 curve in a post-lockdown world [3, 14, 21, 37]. We see these diverse social network data applications, because these data can be naturally represented as a relational network among individuals or groups.

In social network analysis, researchers define the people or entities as *nodes* with various attributes, such as demographic information, where as *edges* represent a paired interaction or relationship between nodes. Facebook data are an example of social network data, where the individuals are the nodes with attributes like hometown, education, and age. The edges are if the individuals are "friends" or not with other individuals.

This natural data structure is also why social network data are very challenging to protect. Imagine our Asian American family is part of a research study to identify successful support systems that facilitate mental health recovery. If a data intruder correctly identified the Asian American family in the social network data, the intruder could potentially discover who else in the data is likely Asian American or have other similar characteristics based on how, when, and at what frequency the family interacts with others.

This example shows how the social network structure itself introduces additional information indirectly, which means less privacy. In other words, social network data provide the necessary information on how people interact with one another, such as developing support groups. But, that network information is what makes it easier for data intruders to violate participants' privacy.

How is the data privacy community addressing this problem? At the time of this book, privacy researchers have only developed a handful of statistical disclosure control methods for social network data. To simplify the problem, some researchers focus on protecting the privacy of the edges or relationships. In other words, researchers fully release information about the individuals and alter whether relationships exist between certain individuals. Under this easier data structure, a few statistical disclosure control methods use local differential privacy on whether an edge should or should not exist between specific nodes in the network.

One local differentially private method assigns a probability of keeping an edge if it exists or does not exist between nodes. For instance, if we wanted to add 5 percent more noise to the network, then the probability of adding an edge when none exists between two nodes is 5 percent. When an edge does exist between two nodes, we would apply the inverse probability, which is a 95 percent probability of keeping the edge.

A drawback to using this method and other similar methods is the relationships across multiple nodes and edges may not be preserved. Imagine that we collected data on departmental collaborations at a university. The data contains a research professor and their students, where each node represented the individual's status (e.g., student) and research topic (e.g., data privacy and confidentiality). In the network, the data show edges between the professor and the students, who all share the same research topic. If we use the local differentially private method I described earlier for this data, each edge would be individually evaluated for privacy protection regardless of the existing edges around it. This means there is a chance the local differentially private method will leave edges between the professor and all but one student. If we were data practitioners examining the data, we would question whether the lack of edge between the professor and the one student was intentional, such as an interesting feature in the data, or noise from the data privacy method.

Not considering the full network of edges is why some privacy researchers prefer to develop algorithms that use the pure differential privacy definition instead of local differential privacy. However, pure differential privacy is much more difficult to achieve. In Chapter 4, we discussed how we must consider the universe of possible versions of the social network data if we want to the use pure differential privacy definition. A possible social network would be starting with a star network, where only one node is connected to all the other nodes. If we were to remove the connected node, then the network becomes an empty network, where there are no edges among all the nodes. This extreme scenario is hard to protect and still only considers how to handle the edges in the network—not node and edge information together.

Privacy experts have proposed some statistical disclosure control methods that protects both nodes and edges, such as modeling the data to create synthetic data (with and without differential privacy). This approach usually results in either not enough privacy protection or data with too much privacy, such as generating a synthetic network dataset that has almost all the nodes connected to one another. For example, my colleagues and I tested various methods of generating differentially private synthetic network data on a college student friendship database. Figure 5.4a shows the

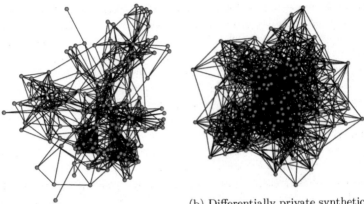

(a) Original friendship data

(b) Differentially private synthetic friendship data

Figure 5.4: The image on the left is the college student friendship data as a social network that contains 162 nodes. The image on the right is the differentially private synthetic data as a social network that resulted in almost all the 162 nodes having an edge or relationship to all the other nodes. Images are from the research conducted by [38].

original social network data, whereas Figure 5.4b displays a denser synthetic social network data by a differentially private method.

To make the situation even more complicated, some social network data have edges that have weights or are directional. Weights represent the strength of the relationship between two nodes and can be given based on several different metrics, such as the number of interactions between two people on social media. On Twitter, if two people retweet or like each other's tweets often, then their relationship weight would be higher compared to someone else they hardly interact with. Email exchanges are an example of directional edges, where an edge would indicate who sent and received an email. Airplane flights are another example of directional network to indicate which airports are the origin or the destination.

With so many open challenges, data privacy research on social network data will continue to grow for years to come and will be driven by the high demand for higher quality data.

5.4   HOW CAN RURAL AMERICA DISAPPEAR?

When state governments issued stay-at-home orders in 2020, companies started providing interactive data visualizations and dashboards to show how well or not so well certain regions of the United States were social distancing. Unacast, a technology company out of New York State, was one of them. They gathered smartphone data from up to 15 percent of people in every county of the United States, and then assigned grades to each state based on how much smartphone users traveled after COVID-19-related closures, compared to before.

In March 2020, news articles, such as *Forbes* and *U.S. News*, praised Washington, DC, Nevada, and other states for reducing their average travel distance by over 50 percent. These regions received an A+, whereas more rural states, such as Wyoming, received an F for only having a 6 percent difference in average travel behavior [26, 36].

At the time, some people made comments that those living in the rural states with low grades were not taking stay-at-home orders seriously and did not care about their communities. As someone who grew up in a rural state and has family in Wyoming, I would argue otherwise—I could see that privacy restrictions on the data led to improper analyses and conclusions.

Given my statement, some might think, "the analysis is the analysis." The data the Unacast analysts collected obviously show that people in rural states are moving the same amount on average as they did before. However, analysts often forget that data analyses are imperfect due to how data are collected. Additionally, the limitations in data collection can create incomplete data stories that are misleading.

But, some might wonder, "What do I mean by having an incomplete or misleading data story? How could the problem be caused by data privacy?" I will use myself as an example again to help explain. Early in 2020, I lived in Washington, DC to start my new position on researching data privacy problems for a public policy research institution. My daily routine consisted of biking five miles from Arlington to downtown Washington, DC and back or taking the metro around the area. At the time, I would stop by the grocery store or at a restaurant for takeout after work each day, where both options were within a half mile radius of my apartment.

When COVID-19 hit the United States and my work closed their downtown office, I stopped commuting to Washington, DC and went to the grocery store once every few days to limit my interactions with people. In this situation, I reduced my average weekly distance of roughly seventy miles down to a mile—more than a 95 percent change in my travel behavior.

When I lived in Salmon, Idaho, my family and I lived five miles outside of the city limits in the country. We also did not grocery shop each day or even every other day. My family went to the local grocery store once a week for fresh produce and made our monthly 140 mile shopping trip to Missoula, Montana to stock up on all the items we needed for the month at Costco and other major stores. In one week, my family and I could easily travel upwards of over 350 miles because of shopping, attending school, or going to work.

Despite a global pandemic, people who live in these rural parts of the country will not change their habit of going to the grocery store once a week or a month. They are already at the minimum frequency to shop for food and other household essentials. Additionally, these trips likely consume the bulk of most rural family's average traveling distance, including mine. This means that even if my family and I were to attend school virtually or work remotely, our overall average distance traveled would still be roughly the same pre- and post-coronavirus. If I remove my commutes to school, then my travel behavior would change from 350 miles to 300 miles, which is approximately a 17 percent difference.

While these two personal scenarios I described are anecdotal, we should expect that for people who reside in Lemhi County (i.e., the county that Salmon is in) are likely to travel more than those who live in Arlington County. Recall my earlier statement that we should be aware of how the data are collected and disseminated. For our example, we should examine the data more closely, such as checking the overall population and county sizes. If we dig deeper, we find that Lemhi County has 7,847 people in 4,569 square miles. In contrast, Arlington's county is 26.1 square miles with a population of 236,842.[6]

Overall, the social distancing dashboard example highlights more problems of not having better access to higher quality data due to privacy concerns for small populations. If Unacast had access to data that tracked people's movement to the census tract or individual level, then the company might have seen the residents of Wyoming practicing social distancing well relative to their essential

---

[6] These population estimates are from the United States Census Bureau for 2019.

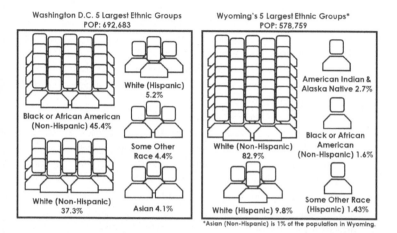

Figure 5.5: Five largest ethnic demographic data of Washington, DC and Wyoming. Estimates are from the United States Census Bureau for 2019. For Washington, DC, the largest ethnic group is Black or African American (Non-Hispanic) at 45.5 percent with Asian as the fifth largest ethnic group at 4.1 percent. For Wyoming, the largest ethnic group is White (Non-Hispanic) at 82.9 percent with Asian not in the top five at 1 percent.

movements, such as shopping at the grocery store. But, as we have learned throughout the book, higher quality information comes at a price of less privacy protection for those who are in the data. Uncast might have lacked more detailed data about the movements because of how most statistical disclosure control methods tend to aggregate or remove information to protect people in smaller sub-populations.

Now, some might think, "If Arlington and Washington, DC have larger populations, then why not report the data down to census tract or block levels for those areas and keep the rural communities at the county-level?" While the question is well intended, it brings up data equity issues that could cause more harm than good.

In this situation, for instance, suppose economists conducted a statistical analysis to develop more targeted economic stimulus plans for supporting small businesses during COVID-19. The analysis would be more accurate on the finer-grained data being down to census tract or block level, understanding the specific needs throughout the county. Arlington County would then receive more precise suggestions to help the residents whereas Lemhi County would receive general support for the 4,569 square mile area, which would likely be less effective. If we extended this example to all counties in the United States, then counties with higher populations would reap more of the economic stimulus package benefits than rural counties.

We can take our theoretical example a step further, where more detailed data can be released if the groups are large enough. Figure 5.5 shows the five largest ethnic demographic groups in Washington, DC and Wyoming, which have similar population totals. Washington, DC has a total population of 692,683 and Wyoming has a total population of 578,759. Based on the ethnic group breakdown, the economists have access to the top five ethnic groups for Washington, DC, and only the top two ethnic groups for Wyoming. They discover through their analysis that small businesses owned by Asian Americans in Washington, DC were hit harder than businesses owned by other ethnic groups. For Wyoming, because the data are generalized to White Non-Hispanic and White Hispanic, they only see that small businesses are struggling.

Suppose our Asian American family of four owned a small restaurant. If they lived in Washington, DC, then they would receive more stimulus money compared to others to keep their business afloat during the pandemic. If our family lived in Wyoming, then they would not receive any additional benefits and may close their business due to Asian American business owners suffering more losses on average. Even if the economists had access to the top five largest ethnic groups in Wyoming, they would still miss any impacts of Asian American owned businesses since Asian Americans are below that threshold. I specifically picked this example, because, in 2020, more Asian American owned businesses closed during the pandemic than others, citing racism and stigmatization as the main factors [15].

Given these problems, we ask ourselves "What is the data privacy community doing now to address the small population problem?" Throughout the book so far, we learned that data curators and privacy researchers have tried many statistical disclosure control methods to preserve the privacy of people in smaller subpopulations, such as data suppression and aggregating the values to larger groups. As we know, these methods are less than ideal because they either withhold the data completely or release suboptimal data for assessing real changes and impacts over time.

On the other hand, if we disaggregated the data or made the small population information available, then we risk fully exposing those in the smaller groups. People of color and people with low-incomes tend to be more vulnerable to data privacy attacks on average than other groups. Also, any attempts to tighten the privacy restrictions on the smaller populations will significantly distort the results, such as early testing of a differentially private method on the 2020 Census [64]. If the data privacy community wants to ensure equitable privacy protections and societal benefits, they should acknowledge and account for these unequal risks before relaxing data protections.

Overall, we must carefully account for how the data are collected and disseminated, or we risk missing a vital piece of the data story we are trying to analyze and understand. Not taking this care can cause unintended harm to those we are trying to protect, such as low income and vulnerable populations, through data privacy policies.

CHAPTER 6

What Data Privacy Laws Exist?

I N 2012, I read *The New York Times* article about Target using statistics to better understand consumers' shopping habits [13]. Like most major retail companies, Target wanted to increase their customer base. But, the company had a particular target in mind—new parents. For this ideal demographic, the retail giant wanted to be the one-stop-shop for groceries, toys, cleaning supplies, and more.

But breaking people's shopping habits is hard. Target's marketing research showed that most consumers buy specific items at specific stores–groceries at a grocery store and toys at a toy store. The research also showed that most people will not break their shopping routine until they experience a major life change, such as having a baby. To beat out the competition, Target wanted to know when a family was expecting well before it became public record.

This is where a statistician and economist, Andrew Pole, came in. Target tasked Pole to analyze Target's massive consumer database to figure out when someone was expecting to have a baby weeks before the due date. And he did. He identified 25 products that indicated if a woman was pregnant and what her estimated due date was. These items were subtle changes, such as buying unscented lotion instead of scented lotion. Pole's prediction model was so accurate that the father of a teenage daughter found out she was pregnant because she received coupons for baby clothes and other baby items.

DOI: 10.1201/9781003122043-6

When the article came out, I had recently graduated with my degrees in mathematics and physics, so I found the article fascinating. I thought, "Wow! How cool that we can harness data to discover unique insights into how humans behave!"

In 2021, I found myself rereading this article and asking myself, "Wow! How would someone feel if they knew their data are being used in this way? How could we allow companies to do this with our personal information?"

Target is not the only company that uses consumer data beyond what people might expect. Think about all the purchase forms you fill, newsletters you subscribe to, and loyalty programs you join. If you provide any personally identifiable information, especially your phone number, then the company that gathered your data will likely be selling it to other entities or be using it in ways that you might not be comfortable with.

Unfortunately, there are no federal data consumer privacy laws in the United States that require companies to tell what they plan to use your data once you agree to surrender your information or force that company to remove that information from their databases. The only data privacy laws are ones internationally and within specific states in America.

In this chapter, I provide an overview of one of the strictest set of consumer privacy laws internationally (the General Data Protection Regulation or GDPR), how it compares against the United States' consumer privacy laws, and what challenges American public policymakers should address to improve the United States privacy laws and learn from GDPR. As I explain these laws, remember that I am providing my perspective as a privacy researcher and that I am not a data privacy law expert. I focus on the parts I think you should be aware of in terms of public policy impacts on society.

For those interested in learning more about United States privacy laws and what could be done to improve them, I highly recommend checking out the Future of Privacy Forum. It is a non-profit organization that "... serve[s] as catalysts for privacy leadership and scholarship, advancing principled data practices in support of emerging technologies."[1]

---

[1] From the Future of Privacy Forum mission statement's at https://fpf.org/about/.

6.1 WHAT IS THE GENERAL DATA PROTECTION REGULATION?

On May 25, 2018, "we updated our privacy policy" notices probably inundated your email inbox from any entity that you might have shared your contact information with. Some might not remember the specific date, but many of us remember the annoyance the flood of emails caused. The reason behind the email intrusion was the General Data Protection Regulation came into effect as European Union law.

The General Data Protection Regulation or GDPR, is a set of European Union data privacy laws and protections. As we learned in previous chapters, how the government and companies collect and disseminate data has significantly evolved over recent decades. This change in processing and analyzing data motivated the European Union to create GDPR. Their old privacy laws, ones from 1995, did not protect people's private information when collected on smartphones.

Some might wonder, "Why did we receive those privacy policy notices? Many of us are not residents of the European Union, and GDPR only protects European Union data." Yes, GDPR specifically protects the personal data of those who live in the European Union. The law also applies to any business or organization that operates within the European Union or gathers personal data on European Union people. This means that many Americans received notices because those organizations, who sent the messages, operated in the European Union or collected European Union personal data. For instance, large corporations that work globally (e.g., Facebook, Amazon, Apple, Microsoft, and Google) must comply with GDPR rules.

Given there are several different types of data privacy, some might ask, "What does GDPR specifically protect?" The 99 articles of GDPR covers a range of topics from what rights does a person have for their personal data to how companies are penalized for violating the law. I will not go over every article of GDPR; that would be another book on its own! I will instead highlight the main parts of GDPR and why it is so powerful in protecting people's personal data.

In the first few articles, GDPR lists the general provisions and definitions. GDPR, for instance, defines to personal data, "...any information relating to an identified or identifiable natural person ([or] data subject)." This information can be a person's name, location data, online identifier, genetic, economic, cultural, social identity, IP addresses[2], and cookies[3].

The data containing identifiers are key for GDPR to consider the information as personal data. Suppose we survey a community for their thoughts on creating a new dog park. If we collect people's names or email addresses, the data would be subjected to GDPR rules, whereas GDPR would not apply if we keep the survey anonymous.

Another distinction for GDPR is it also separates personal data into two categories: general personal data and sensitive personal data. This European Union law classifies sensitive data as racial or ethnic origin, political opinions, religious or philosophical beliefs, trade union membership, the processing of genetic data and biometric data for the purpose of uniquely identifying a natural person, and sex life or sexual orientation. Revealing this type of personal data is prohibited. GDPR precludes the processing of sensitive personal information and only for special circumstances can a company process such data.

For those responsible for protecting personal data, GDPR defines them as either controllers or processors. The *controller* is the one responsible for deciding the purpose and processing of the personal data, whereas the *processor* acts on behalf of the controller while processing personal data. A classic example is a doctor's office uses an automated system to notify patients in the waiting room to proceed to an examination room. The system contains both audio and visual indicators of the patient's name and which room to enter. In this case, the doctor would be the controller of the personal information, who determined how the data would be processed by the waiting room notification system. Another example is a museum hires a printing company to make and send invitations for their upcoming fundraiser event. The museum would be the controller of the personal data, which requested the printing company to process the information to create and send the invitations.

However, the roles may change if the person or entity acts within their professorial obligations. Imagine a company hires an

---

[2]IP or Internet Protocol address is an identification number that is assigned to every device that connects to a computer network.

[3]Cookies are small pieces of data generated when a person visits a website and that information is stored by their web browser. Although cookies do not contain any information that identifies the person, the information generated by cookies could be linked to other sources or be used to track people as they navigate through different websites.

accountant to balance and monitor their budget. Some might think the company as the controller and the accountant as the processor. Rather, the accountant becomes the controller due to their professional responsibility on how they process the data. For instance, the accountant may discover malpractice within the budgets and decides to report it to the police. The accountant did not act on behalf of the company that hired them, becoming a controller. In other words, any person or entity that processes data within their professional duties will always be the controller, because they cannot "hand over" or "share" their controller responsibilities to the company.

After provisions and definitions, GDPR states the seven principles for processing personal data. These seven principles are called:

1. Lawfulness, Fairness, and Transparency;

2. Purpose Limitation;

3. Data Minimization;

4. Accuracy;

5. Storage Limitation;

6. Integrity and Confidentiality; and

7. Accountability.

They guide the remaining articles in GDPR and aim to give every person in the European Union "the right to privacy and the right to be forgotten." Specifically, GDPR monitors what organizations do with people's personal data, empowers people to control how their personal data are being collected and used by others, and forces organizations to justify *anything* they do with people's personal data.

GDPR addresses this last goal in Article 6, which only allows organizations to use personal data if at least one of the six lawful bases apply. These bases are referred as:

1. consent,

2. performance of a contract,

3. legal obligation,

4. vital interest,

5. public interest, and

6. legitimate interest.

I will describe each one.

GDPR explicitly states that *consent* must be "...freely given, specific, informed and unambiguous indication of the data subject's wishes by which he or she, by a statement or by a clear affirmative action, signifies agreement to the processing of personal data relating to him or her." This means the company must ensure the person understands clearly why their personal data are collected and what the data will be used for. Acceptance to the use of someone's personal data needs be separated from other terms and conditions the company might have. Essentially, businesses cannot use someone's personal data without someone's *explicit consent* that is freely given and informed.

An example is when companies ask whether we want to be on their newsletter email list after purchasing an item on their online store. Under GDPR, that checkbox needs to be unchecked by default. The company must then provide information on how the personal data will be collected and used if the person *opts-in*. This is the opposite of what is done in the United States, where the checkbox is typically checked and people must *opt-out* to not have their personal data collected. Additionally, United States businesses are not required to explain how that information will be used beyond sending newsletters.

For *contract*, a processor requires someone's personal data to complete specific steps outlined in a contract or prior to entering into a contract. Simple examples include buying a house or car, where the business must use names and addresses to mail the house or car title.

Legal obligation is when a company must use personal data to comply with other laws. In the United States, all organizations with unemployment insurance must report employee data to the Bureau of Labor Statistics per United States law.

Under *vital interest*, a business can access personal data to protect the person's life and others. Suppose a person has an accident at work. That person's company can then use personal data to report any specific health related issues to the hospital the company may have on file.

For *public interest*, an organization might perform certain tasks for people's public interest or to complete official functions. For instance, many local governments will collect personal data to identify what public goods are needed for the city, such as building parks or libraries.

A company may use personal data for *legitimate interests*. These interests may include situations where someone would expect an organization to collect personal data, use the data in ways that are low-risk, or make a compelling argument on the benefits outweighing the risks. Imagine someone falls ill or is hurt at work. If the company collected emergency contact information, then the company could use it to notify the injured person's listed contact. However, the person may object to having their data collected for the legitimate interest, because GDPR requires full transparency on how and why the personal data are gathered.

We next focus on Article 25 or "data protection by design and by default." It states that any organization that plans to collect, store, and disseminate personal data must use the available "state of the art" privacy preserving methodology and technology. The organization's resources, context, and scope for the data process determines the extent of what "state of the art" methodology and technology is implemented.

For publicly releasing data, Article 25 requires businesses to apply certain statistical disclosure control methods that account for people's privacy rights. This means that an organization must take this limitation into account as they assess how they will use personal data. This assessment is a part of the data privacy workflow we covered in Chapter 4, Section 4.4.

Given how strict GDPR is, some might wonder, "What discourages a company from violating GDPR?" Unlike some United States environmental laws, where the penalty tends to be cheaper than fixing the problem, GDPR sets fines to make non-compliance costly for both small and large businesses. GDPR accomplishes this through a two tiered administrative fine system in Article 83.

Some classify the first tier as *less severe infringements*, where these fines are subject to administrative fines up €10 million,[4] or up to 2% of the total worldwide annual turnover of the preceding financial year, whichever is higher. Less severe infringements are violations that pertain to controllers and processors, certification bodies (i.e., accredited entities that verify if an organization is compliant to GDPR law and must be unbiased and transparent of their evaluation), and monitoring bodies (i.e., designated entities

---

[4]At the time this book's publication, this amount converts to roughly a little over $12.2 million in US currency.

Table 6.1: The currency conversions for the General Data Protection Regulation fines in millions. The table displays the fine in the original currency and the converted values to past (when the fine occurred) and current (*fine at the time of the book publication) United States (US) currency values.

COMPANY	FINE	PAST US	CURRENT US*
Google	€50	$57	$61
H&M	€35.3	$41.3	$43
British Airways (reduced fine)	£20	$25.9	$28.3
British Airways (original fine)	£183	$236.9	$259

with the expertise to independently handle complaints and report infringements in an unbiased manner).

The next tier is for more *serious infringements*, where the company breaches the main goal of GDPR, "the right to privacy and the right to be forgotten." In other words, these infringements violate the basic principles, conditions of consent, the person's right to know how their personal data are being collected and used, and the transferring of personal data to another country or international organization. Businesses that violate GDPR in these ways are subjected to administrative fines up to €20 million,[5] or up to 4% of the total worldwide annual turnover of the preceding financial year, whichever is higher.

Since GDPR became law in 2018, the European Union has enforced a total of over $330 million in fines. One of the biggest offenders was Google in 2019 for not properly informing people how their personal data were being collected and used for personalized advertisements on Google's search engine, Google Maps, and YouTube [51]. The European Union fined the tech giant €50 million for the violation. In October 2020, the Data Protection Authority of Hamburg fined H&M, a clothing company, €35.3 million for collecting employees personal data since 2014 and creating personal profiles of them, such as details about their vacation trips, religious beliefs, medical conditions, and more [47].

---

[5]At the time this book's publication, this amount converts to roughly a little over $24.3 million in US currency.

Although we did not go over the GDPR articles that include data breaches, we know that Article 25 requires companies to use the state of the art in technology to protect people's data. In 2020, the United Kingdom Information Commissioner's Office fined British Airways £20 million for a major personal and credit card data breach. British Airways lacked sufficient data security measures they could have implemented through their Microsoft operating systems, such as multi-factor authentication. The Information Commissioner's Office originally planned to fine £183 million, but the fine was significantly reduced due to British Airways updating their security systems and fully cooperating with the investigation [61].

6.2 WHAT ARE THE CHALLENGES FOR THE GENERAL DATA PROTECTION REGULATION?

Although GDPR is "leading the charge" in improving people's data privacy, the European Union laws have several flaws to overcome. I will only focus on a few of these challenges.

The biggest complaint against GDPR is the cost of compliance. According to *Forbes*, in 2018, the Fortune 500 and United Kingdom Financial Times Stock Exchange 350 companies spent $7.8 billion and $1.1 billion, respectively, to be GDPR compliant [56]. The initial starting expenses ranged from upgrading technology to hiring more employees, such as lawyers and privacy experts. Additionally, companies must continue investing to maintain staff training, monitor compliance, and update their technology over time. Many argue the investment is a necessary part of the process to protect people's personal data.

As a counterpoint, some economists contend not every organization has enough funding or the basic infrastructure to easily update their privacy and security systems [35]. These costs have a huge impact on smaller businesses that often struggle to be GDPR compliant given the complexity of the law and lacking the initial personnel to help with compliance. In contrast, tech giants tend to already have the personnel and technological infrastructure and, if necessary, the capital to hire more staff or to upgrade their infrastructure.

As a consequence to avoid the costly upgrades and non-compliance penalties, many small- to medium-size companies[6] have

---

[6]According to the European Commission, small businesses have a staff

either withheld from the doing business within the European Union or destroyed the personal data they have collected.

For example of businesses withdrawing from the European Union, Pottery Barn, an American furnishing retailer, stopped shipping to European Union addresses [31]. Additionally, over 1,000 United States news sites decided to block European visitors [58]. J.D. Wetherspoon, a British pub chain, opted to delete its entire email database that once contained over 650,000 emails [40].

When smaller companies departed the European Union market or foregone their personal data assets, they created a void that allowed larger corporations to strengthen their dominance in their respective markets. Politico reported that Amazon, Facebook, and Google increased their online advertising market share within the European Union after GDPR came into effect [53].

This situation brings up the question, "How much does it cost a smaller company to become GDPR compliant?" In 2019, GDPR.EU, a GDPR resource website operated by Proton Technologies AG, conducted a survey on 716 small business throughout Spain, the United Kingdom, France, and Ireland. These small businesses had fewer than 500 employees and must follow GDPR rules. GDPR.EU found that just over 60 percent of the small businesses surveyed paid between €1,000 to €99,999 (i.e., about $1,200 to $122,000) and 17 percent paid between €100,000 to €999,999 (i.e., about $122,000 to $1.22 million) to be GDPR compliant. For these businesses, the most common expenses came from training employees, hiring consultants, and purchasing software and equipment [19].

Another issue is several organizations will likely eliminate flourishing business models or cancel research and development in areas that might be too difficult to ensure GDPR compliance. Many tech startups, for example, use a free or freemium model, a business pricing strategy where the basic product is free and the customer can spend money to access additional services. This business model often depends on collecting personal data to gather revenue

---

count of less than 50 with either a turnover or a balance sheet total that is less than or equal to €10 million, which is roughly $12.2 million in US currency. Medium-sized businesses have a staff count of less than 250 with either a turnover that is less than or equal to €50 million or a balance sheet total that is less than or equal to €43 million, which are roughly $61 million and $52.4 million in US currency, respectively. In the United States, the Small Business Administration sets the standards for small business size by industry. A small business in the manufacturing industry, for example, has fewer than 500 employees and has less than $7.5 million in revenue.

through advertisements. We commonly see this model when accessing free news article with advertisements in between paragraphs or online gaming with advertisement pop-ups.

Other businesses will offer personalized advertising or recommendations, such as Amazon suggesting a certain product over another or Spotify and Pandora creating tailored radio stations. Most of these services will have to develop a new business model or eliminate further development due to GDPR restrictions.

Some might argue, "What if the data are used for the common good? Could the end justify the means?" The examples we have covered so far are for-profit institutions, but there are other organizations that apply personal data for the public good in inventive ways.

For example, DataKind is a data science organization that harnesses data in collaboration with other groups "...to tackle critical humanitarian issues in the fields of education, poverty, health, human rights, the environment and cities." Their projects include creating safer streets, forecasting water demand, and finding paths out of homelessness. In the United States, Data for Black Lives uses data science "...to create concrete and measurable change in the lives of Black people," such as creating an interactive map that allows the user to explore the disproportionate impact of COVID-19 on black people in the United States. Another example is a startup called, Flowaste, that uses data to reduce food waste, starting in cafeterias.

Although these organizations are non-profits using data for the public good, GDPR is applied equally for all organizations. This means non-profits must also be aware of how they use personal data to be GDPR compliant, which can take away precious time and money from other potential initiatives.

Some might then wonder, "What if the companies could obtain consent from each person to use their personal information? If so, could the services or projects the companies conduct still continue?" GDPR allows the use of personal data if consent is given. However, an unintended consequence is that most recommendation algorithms or services that organizations developed in the past rely on data collected over several years prior to GDPR. For these businesses to use the data, they would have to receive consent from every single person who contributed to that data throughout the years. The hassle of doing so is why several organizations decided to destroy their data and start over or remove themselves from interacting with customers in the European Union.

Additionally, issues of consent escalate in other ways GDPR tried to prevent, but did not succeed—the unequal power between employers and employees. Employers could pressure employees to consent to their personal data being collected that is outside of the other five lawful bases. The employers could say, "If you want to keep your job, then you must grant your consent for your personal data being gathered by the company."

6.3 WHAT DATA PRIVACY LAWS EXIST IN THE UNITED STATES?

As stated in Chapter 1, the United States has no federal law that regulates how personal data can be collected, stored, and used. What laws exist on data collection and dissemination are restricted specific to federal agencies, such as the United States Census Bureau and Internal Revenue Service.

At the state level, California, Maine, and Nevada are the only states with privacy laws to protect peoples' privacy. At the time of this book's publication, eleven states (e.g., Virginia and Washington) have active bills, but have not completed the legislative process while Mississippi and North Dakota failed to pass the bill after being introduced.

Out of the three states that passed privacy laws, the California laws are by far the strongest and most comprehensive privacy regulations in the United States. On June 28, 2018, Jerry Brown, the Governor of California at the time, signed California Consumer Privacy Act into law, which became effective on January 1, 2020. Then on November 3, 2020, the majority of California residents voted yes to Proposition 24 or the California Privacy Rights Act that builds on and strengthens California Consumer Privacy Act and will be enacted on January 1, 2023.

For the sake of brevity, I will refer to these acts as the California Privacy Act and its amendment.

Since these laws are applied to the state level, some might ask "Does the California Privacy Act and its amendment affect the rest of the United States or does it not?" Many policymakers argue that a federal law is needed for having stronger privacy laws, such as enforcing harsher penalties for being non-compliant. Yet, California has the most influence out of all the states due to its economic power.

Imagine if California became a country. The state would be the fifth largest economy in the world based on gross domestic product.[7] Additionally, Silicon Valley calls California home, which means that many of the tech giants that are headquartered there must comply with the California Privacy Act and its amendment. This is why many received another wave of privacy policy notices on January 1, 2020.

Does this mean the California Privacy Act and its amendment are equivalent to GDPR, but for California? Not quite. While these data privacy laws share similar goals to provide people more control over their personal data, the California Privacy Act and its amendment apply to a smaller subset of businesses than GDPR and significantly differs on several other aspects. A few of these differences make the California Privacy Act a weaker set of laws than GDPR. This situation is why many public policymakers initially referred to the California Privacy Act as "GDPR-lite" and why lawmakers amended the act.

One of the biggest differences between the California Privacy Act and GDPR is which organizations must comply with these laws. We learned that GDPR regulates all businesses that operate within the European Union or gathers personal data on European Union people. The California Privacy Act instead specifies a business as a for-profit company that collects consumers' (i.e., a natural person who is a California resident) personal information (e.g., includes on behalf or jointly with another entity), determines the purpose and means of processing that personal information, does business in the State of California, and satisfies one or more of the following criteria:

1. An annual gross revenue in excess of $25 million.

2. Alone or in combination, annually buys, receives for the business's commercial purposes, sells, or shares for commercial purposes, alone or in combination, the personal information of 50,000 or more consumers, households, or devices.

3. Derives 50 percent or more of its annual revenues from selling consumers' personal information.

---

[7]For 2019, the United States Bureau of Economic Analysis reported that California had a gross domestic product of about $3.2 trillion. Only the rest of the United States, China, Japan, and Germany have larger gross domestic products.

California sales, property or payroll

Year	CA sales exceed (either the threshold amount or 25% of total sales)	CA real and tangible personal property exceed (either the threshold amount or 25% of total property)	CA payroll compensation exceeds (either the threshold amount or 25% of total payroll)
2020	$610,395	$61,040	$61,040
2019	$601,967	$60,197	$60,197
2018	$583,867	$58,387	$58,387
2017	$561,951	$56,195	$56,195
2016	$547,711	$54,771	$54,771
2015	$536,446	$53,644	$53,644
2014	$529,562	$52,956	$52,956

Figure 6.1: A screenshot of the "California sales, property or payroll" table on the California Franchise Tax Board website at https://www.ftb.ca.gov/file/business/doing-business-in-california.html.

Additionally, the California Privacy Act requires any organization that controls or is controlled by a business that satisfies the California Privacy Act definition to comply.

Given this definition, does this mean the California Privacy Act extend beyond California residents? Unlike GDPR, the California Privacy Act is less clear on whether "does business in the State of California" applies to organizations out-of-state. To clarify, the State of California Franchise Tax Board considers "doing business in California"[8] as a business that engages in any transaction for financial gain in California, is organized or commercially domiciled in California, and/or has California sales, property, or payroll exceed the amounts listed in Figure 6.1.

Despite these constraints, the California Privacy Act applies to roughly half a million United States companies.[9]

---

[8] Information on "Doing Business in California" can be found on the State of California Franchise Tax Board website at https://www.ftb.ca.gov/file/business/doing-business-in-california.html

[9] Statistic obtained from the International Association of Privacy at https://iapp.org/news/a/new-california-privacy-law-to-affect-more-than-half-a-million-us-companies/.

Another important distinction between the California Privacy Act and GDPR is how they define personal information or personal data. The California Privacy Act uses the term personal information, whereas GDPR uses personal data. The California Privacy Act and GDPR define these terms similarly, so I will use them interchangeably. But, a major deviation in how the California Privacy Act and GDPR refer to personal information is scope. The California Privacy Act does not protect personal information that federal, state, or local governments make publicly available lawfully, such as the Bureau of Labor Statistics reporting unemployment rates in California. In contrast, GDPR requires *any entity* that collects personal data from public data to follow GDPR laws.

The California Privacy Act also does not have a special category that is equivalent to GDPR's sensitive personal data. In particular, GDPR classifies data concerning health, genetic data, and biometric data as sensitive personal data, which protects the data under a separate article. The California Privacy Act does not directly protect medical data, because the United States has the Confidentiality of Medical Information Act.

What legal grounds businesses may use personal information is another area the California Privacy Act significantly deviates from GDPR. We learned earlier in the chapter that GDPR has the six lawful bases for using personal data, which forces companies to justify using the data before data collection begins. For the California Privacy Act, the legal grounds come into play during or after companies collect data.

In other words, the California Privacy Act requires the organizations that collect personal data to inform consumers what is being collected and allow those consumers to opt-out of the data collection. Companies must then obtain explicit consent to sell or disclose any of that information and provide a process for people to request all that information to be deleted. However, the California Privacy Act has no limitations on the data collection, purpose, or storage, whereas GDPR does.

Does the California Privacy Act also deviate in how it penalizes businesses for non-compliance? The California Privacy Act specifies the Attorney General of California may apply a civil penalty of \$2,500 per violation or \$7,500 per intentional violation to the offending company if not addressed within 30 days of the violation. While this penalty seems low, the California Privacy Act sets no cap for the number of violations that can be charged, so the total amount could be quite large.

Facebook, for example, has roughly 223 million United States users in 2020.[10] Suppose we find violation for 0.1 percent of those users (i.e., 223,000 users). Facebook would be charged a civil penalty as little as $557.5 million or as large as $1.7 billion, depending on the type of violation. Both theoretical civil penalties would be larger than any GDPR fine at the time this book's publication.

Knowing the California Privacy Act weaknesses, some might wonder, "How does the amendment strength the California Privacy Act?" In general, the amendment expands existing consumer rights, creates new consumer rights, increases requirements on businesses, and more. We will not cover all 31 Sections of the amendment. Rather, we will focus on a few of the changes that transformed and strengthened the California Privacy Act.

Originally, the California Privacy Act gave people the right to opt-out the "sale" of their personal information. In Section 8, the amendment expanded this right to opt-out to both the "sale" and "sharing" of consumers' personal data. The amendment defines "share," "shared," or "sharing" as "...renting, releasing, disclosing, disseminating, making available, transferring, or otherwise communicating orally, in writing, or by electronic or other means" to a third party that does any cross-context behavioral advertising and whether or not the advertising is used for monetary gain.

This clarification addresses how businesses operate online advertising and provides consumers the right to delete their personal data. Businesses must also "...notify all third parties to whom the business has sold or shared such personal information, to delete the consumer's personal information, unless this proves impossible or involves disproportionate effort." Furthermore, in Section 14, the amendment updated what qualifies as a "business" by changing the threshold amount of personal information that a business can have from 50,000 to 100,000.

Another expanded consumer right is the right to access personal information indefinitely. The California Privacy Act originally only allowed consumers to request personal information from the previous 12-month time frame. Under Section 12 of the amendment, "...the business shall be required to provide such information unless doing so proves impossible or would involve a disproportionate

---

[10]Information obtained from Statista, a German based company that specializes in consumer and market data.

effort." This requirement only applies to personal data collected on or after January 1, 2022.

In addition to expanding rights, the amendment created new rights. The amendment added Section 6, the right to correction of inaccurate information. When a consumer makes a valid request to correct inaccurate personal information, the business "...shall use commercially reasonable efforts to correct the inaccurate personal information, as directed by the consumer."

The amendment also added Section 10, where a consumer can request businesses to stop using or disclosing sensitive personal information. This section came with a new category, sensitive personal information. The amendment considers sensitive data similarly to GDPR and broadens the definition to include government-issued information, account log-in credentials, financial account information, precise geolocation, and information in various forms of messaging.

Another similarity to GDPR is the amendment established purpose limitation, data minimization, and storage limitation. Businesses can only collect, use, retain, and share personal information for what is "reasonably necessary and proportionate" to achieve the businesses' purposes or related contexts. For data storage, the business cannot collect personal data "...for longer than is reasonably necessary for that disclosed purpose."

To enforce these laws, the amendment established a new enforcement agency, the California Privacy Protection Agency. Section 24 outlines that this agency will be governed by five members with expertise in privacy, technology, and consumer rights. Additionally, the California Privacy Protection Agency will be responsible for enforcing the California Privacy Act and its amendment, raising public awareness, and guiding businesses in best practices. With a new enforcement entity, the fine changed from being a civil penalty to an administrative fine. Lastly, the amendment eliminated the 30 day window for correcting violations and defaulted violations involving a person younger than 16 to the $7,500 fine.

6.4 WHAT ARE THE CHALLENGES FOR FUTURE UNITED STATES DATA PRIVACY LAWS?

Although the California Privacy Act and its amendment provide stronger privacy laws, the United States still needs data privacy laws at the federal level. Before that happens, we should learn from GDPR and the amended California Privacy Act to avoid the challenges these current laws face.

While the cost of compliance is one of the challenges we covered for GDPR, the amended California Privacy Act avoids this issue by requiring only larger businesses to comply. However, policy researchers argue that not requiring all United States organizations to comply with privacy laws creates a hole in our country's privacy protection [29]. A possible solution is to create a set of scaled compliance requirements based on several criteria, such as the company size and the type of data it collects, uses, and stores [30].

Another challenge is having opt-in instead of opt-out privacy laws. Many of us have likely fallen victim to a pre-ticked box that subscribed us to a business' mailing list, causing lots of frustration. The amended California Privacy Act created the right for people to opt-out of their personal data being used, but the default should be people opting-in to have certain services.

Yet, several conveniences many of us enjoy will potentially either drop in quality or cease to exist under an opt-in federal privacy law, because people are less likely to opt-in versus opt-out. For instance, Google Maps relies on crowdsourcing people's geospatial location data to warn drivers when there are traffic delays, unexpected construction, or car crashes. Another example is recommendation systems that help you discover new music or shows, such as Spotify and Netflix, which depend on millions of people's profiles on their "likes and dislikes." Web search engines use a large number of inquiries to suggest keywords or phrases, provide the top web pages to browse, and recommend other related searches.

One of the harder challenges for both GDPR and the amended California Privacy Act is empowering workers' privacy rights. These rights must be better established, especially as more people will likely continue to remote work after the pandemic. More companies are adopting surveillance technology to monitor worker attentiveness, track what website the worker visits, record keystrokes, and more. The companies then use this information to conduct performance analytics to assess employee productivity, such as how much time they are spending on certain tasks.

Similar performance analyses are being done in colleges. For instance, when I spoke with some of my friends, who teach in higher education, many commented that newer university dashboards show when students are accessing homework assignments or other course materials. These interfaces display the students' personal data as data visualizations or summary statistics to inform professors how to improve the learning experience for their students. While student performance information is helpful, my colleagues expressed concern about how invasive it felt to them learning their students' personal habits, such as receiving notifications that a student often logs in at 11 pm.

The intrusion on students' privacy does not end here. Some universities track students via a smartphone app to know when students are late, absent, or left class early [24]. Normally, this data collection would violate GDPR and the amended California Privacy Act. But, a university could require students to consent to their personal data being collected if they want to attend school.

As possible solutions, some researchers suggest creating clear guidelines and rules for remote working and learning, setting expiration time on storing data, and establishing specific laws to address "forced consent" when there is an imbalance of power between an entity and a person [41, 63].

Given these ongoing challenges, we clearly need better communication and educational materials on the interaction between privacy laws and technology. If we develop these materials, privacy researchers could more easily communicate and work with public policymakers to make more informed public policy changes.

CHAPTER 7

What Is the Future of Data Privacy?

A N IMPORTANT ROLE FOR A SCIENTIST is disseminating their work and tailoring that communication to their target audience. The research is meaningless if the researcher cannot explain it to others—no matter how groundbreaking it is. Scientists who cannot properly describe their research will have fewer people use it, resulting in less scientific or social impact. Furthermore, they will have a harder time winning grants to fund their research.

To help spread their knowledge, researchers often attend science conferences. These formal gatherings provide a platform for scientists to present their research findings, learn from other scientists, and network to form collaborations.

Several years ago, I traveled with a friend to a statistics conference. She specialized in a different research area than me, so we talked about each other's research. At some point in our conversation, my colleague asked, "What do you think about the future of our data privacy?"

When I answered, I blurted out, "I think *our* data privacy is screwed."

My friend snapped back about why I should bother researching data privacy if I thought people's privacy was "skrewed." Based on her response, I could tell she was taken aback by my comment because my initial answer did not fully convey my intent. I failed to explain my research to my target audience—another researcher and friend.

DOI: 10.1201/9781003122043-7

My simple response needed more explanation for my friend. When I said *our* in the conversation, I referred to the millennial and older generations. We experienced the explosion of the internet and search engines, the birth of social media and how it took the world by storm, and the convenience of pocket computers (i.e., smartphones). Society learned the hard way about how much data technology collects and disseminates.

For these reasons, I told my friend our society had already lost a lot of our personal privacy. However, we can still protect our future personal information with better data privacy methods and laws. I gave this explanation to her before emphasizing that my motivation is to protect our future data and the next generations' information thereafter.

Now, my motivation is tackling data privacy challenges within public policy and developing better ways of communicating data privacy research. To close this book, I discuss what I identify as the important outstanding data privacy challenges and what I recommend the data privacy community do to help protect people's privacy for the public good.[1]

7.1 WHY ARE THERE NOT ENOUGH USE CASES?

The adage that "practice makes perfect" holds true for many things, including creating new statistical disclosure control methods. The data privacy community have developed and tested traditional statistical disclosure control approaches for *several decades*, whereas differentially private methods have existed for 15 years as of the book's publication.

Most of the differential privacy research are also *still* theoretical. The early practical studies mostly focused on complex data, which faced considerable resistance from many data users. If we want a wider acceptance of differential privacy, we need more small, practical differentially private applications to discern what are some of the data challenges and how we should address them.

In 2020, Google quickly developed their mobility reports to provide access to invaluable data during a global pandemic— likely building from their past experiences with implementing various differentially private applications [1]. LinkedIn also revealed their LinkedIn's programming interface that protected LinkedIn

---

[1]I co-authored a non-technical article meant for statisticians that covers similar material for a statistics magazine called, *How Statisticians Should Grapple with Privacy in a Changing Data Landscape* [57].

members' engagement data [49]. Both companies developed their differentially private methods for specific applications that contributed to their success.

Another example is the Bureau of Labor Statistics' Quarterly Census of Employment and Wages data from Chapter 4. My colleagues and I implemented differentially private methods on a small data as a pilot study, and we learned quickly which differentially private methods performed better than others. This information allowed us to narrow the methods for larger scale testing, saving us time. We also identified unforeseen issues that we would have likely missed if we immediately started with the full data [7].

Unlike differential privacy, traditional statistical disclosure control methods have the opposite problem. Although privacy researchers and data practitioners have a solid foundation to build from, they have little experience with applying traditional statistical disclosure control methods to complex data.

From Chapter 4, researchers generated synthetic data on the Internal Revenue Service's non-filer tax data. The data had nineteen variables, whereas the individual income taxpayer dataset contains 200 variables. These variables also have more complicated constraints, such as whether a record qualifies for charitable deductions. Despite these additional challenges, the researchers who worked on the non-filer tax data already know which synthetic data method to start with and have more experience on how to handle the more difficult constraints.

Similarly, Maryland Longitudinal Data System Center researchers started with a well established statistical disclosure control method for their longitudinal education data. They then iterated through the workflow in Chapter 4 to refine and expand their method to handle thousands of variables [20].

The privacy researchers in these examples partnered with subject matter experts (e.g., tax policy experts on the income tax data) to identify the key parts of the confidential data to preserve. The privacy researchers do not know the data as well as the subject matter experts, who also have a different perspective that could provide alternative solutions. These collaborations will be essential in conducting more useful and practical applications.

7.2 WHY USE A TIERED SYSTEM TO ACCESS DATA?

Throughout the book, we learned the different ways we can access data, ranging from publicly released data to the Federal

Statistical Research Data Centers. Recently, synthetic data generation has grown significantly in popularity as the leading statistical disclosure control method with and without differential privacy.

Despite the method's popularity and several successful applications, synthetic data generation is not enough on its own to address all the researchers' and data users' analyses. This situation returns us to our original discussions in Chapters 3 and 4 on the trade-off between data privacy and data utility. Even if we selected an appropriate model, the synthetic data might lose the interesting features for key public policy features. Additionally, data privacy research has shown synthetic data will often not provide accurate estimates for more complex statistical models.

Some might ask, "What are future possible solutions to increase data access while preserving some privacy?" Synthetic Data Server in Chapter 5, Section 5.2 hosts two synthetic datasets for the United States Census Bureau. After researchers and data practitioners test their statistical analyses on the synthetic data, they can then submit the same analyses for approved staff members, such as federal workers, to validate if the synthetic data results are comparable to the confidential data results. The data privacy community refers to this process as a validation server. Many privacy researchers like the validation server idea because it allows researchers and data users to run analyses on the confidential data without see it.

However, the validation server described here has two major drawbacks. First, because the validation server is not automated, the server consumes a lot of limited staff time. Additionally, the demand often exceeds available staff resources, causing even longer delays for approved results to return to the researcher. Second, staff manually vets each statistical release for disclosure risks. This process tends to be labor-intensive and imperfect since it relies on subjective human review.

Both disadvantages cause federal agencies to limit the number of people who can gain access to this system, increasing the already tremendous demand from researchers and data practitioners. This demand indicates that much more research could be conducted if a safe and less resource-intensive process were developed to expand data access.

One way to alleviate the burden on staff members' time and other resources is to automate the validation server. Members of the data privacy community are working on creating an automated validation server that may use a version of differential privacy to formally constrain results from submitted analyses. Developing such a system is "easier said than done." There are multiple challenges that includes how do we implement a practical and appropriate privacy loss budget for researchers and institutions, how do we inform the trade-offs of privacy and utility properly to users, how do we educate users on interpreting results that are altered or noisy, how do we develop safeguards to prevent users from "gaming the system," and more.

Again, there is no "one-size-fits-all" when it comes to gaining data access. In some cases, researchers only need the synthetic data while others must apply their more complex analyses to the confidential data. With this range of data access needs, some in the data privacy community are working towards a tiered system, such as the following:

1. **Basic Access:** Anyone can access a publicly released dataset, which will most likely take the form of synthetic data.

2. **Validation Server Access:** Researchers and data users must first test their statistical analyses on the synthetic data and then submit an application to request access to the validation server. Once approved, the researchers and users may send their statistical analyses through the validation server online interface.

3. **Full Access:** Researchers and data practitioners must obtain full clearance to access the confidential data.

The Basic and Full Access is what currently exists for most federal agencies, whereas the Validation Access is a newer idea. As an extra security measure, Validation Server Access requires an application that would ensure the researcher thought through their research questions and tested their proposed analyses on the synthetic data. The application should also help staff assess if the applicant is trustworthy without being too time consuming to review.

Some envision the tiered access system as a potential piece or road map for creating the National Secure Data Service described in Chapter 2, Section 2.3. The development and deployment of the National Secure Data Service and tiered access system is another step forward to enable more people to safely access confidential data. But, this system is a large undertaking. It will require multiple organizations to come together to make it a success along with changing the laws, such as amending Title 13. Additionally, the data privacy community must carefully construct and rigorously test the system to avoid exacerbating some issues, such as the inequality of accessing the data. Given the size and complexity of implementing a tiered access system, there will likely be unforeseen challenges that are more difficult than the ones already identified.

7.3 WHAT CAN BE DONE TO ADDRESS THE INEQUALITY IN DATA PRIVACY?

During the summer of 2020, the social unrest and pandemic motivated my colleague and I to draft a short blog[2] about the privacy inequality for underrepresented groups. I covered some topics the blog discusses in this book, such as how the current data privacy approaches tend to harm minorities rather than help and how COVID-19 will make the inequality issue worse.

However, we realized we never suggested any potential solutions to the problem. My colleague and I brainstormed some ideas, but the data privacy inequality issue is *hard*. Neither us nor the data privacy community have researched the area much—if at all. We could only come up with privacy experts must *start researching more on data equity issues.*

Despite the importance, most data privacy research does not include or address data equity problems. This situation results in little to no proposed solutions as discussed in Chapter 5. Given the lack of possible research directions, I initially thought about not including the data privacy inequality problem as a subsection in this final chapter. But, as I wrote this book, all I could think was, "We must start somewhere."

In early 2021, the nation would be forced to *start somewhere.* On his first day in office, the 46th United States President, Joseph Biden, signed an executive order[3] that aims to "...advanc[e] equity for all, including people of color and others who have been historically underserved, marginalized, and adversely affected by persistent poverty and inequality." This executive order instructed federal agencies and White House offices to investigate the "barriers to equal opportunity" across the federal government by conducting

---

[2]The blog is titled, *Another Invisible Threat: Unequal Privacy Cost of Tracing COVID-19* and is currently unpublished.

[3]Biden signed the *Executive Order On Advancing Racial Equity and Support for Underserved Communities Through the Federal Government* on January 20, 2021.

an equity assessment. Additionally, the executive order highlights that "many Federal datasets are not disaggregated by race, ethnicity, gender, disability, income, veteran status, or other key demographic variables" and the "lack of data has cascading effects and impedes efforts to measure and advance equity." Furthermore, the executive order established the Equitable Data Working Group, an interagency group that reports to the President and evaluates the current conditions of federal data infrastructure and provide recommendations for improvements.

Now, some might wonder, "How will this executive order address data equality? What does it have to do with data privacy?" Although the Centers for Disease Control and Prevention declared COVID-19 hits communities of color the hardest, these communities are not receiving the aid they need [10]. Health officials say the lack of disaggregated data by race and ethnicity affected timely vaccine distribution and proper messaging for correcting misinformation to help those communities [42]. The executive order will allow federal agencies and policymakers to publicly release this data.

However, we know that federal agencies must carefully consider the harms they could cause to the people they are trying to help when releasing more detailed data. Then, another question becomes, "What should we do to balance these competing needs?"

A possible solution is we should first build on past and current technological innovations when expanding access to highly confidential data, such as synthetic data generation and differential privacy. Many federal agencies are still using older statistical disclosure control methods that either make the data too private (e.g., lots of suppression) or not private enough (e.g., not altering or adding enough noise to protect records). Some of these agencies want to implement modern statistical disclosure control methods on their own data, but lack the educational materials and computational resources to do so. This means the data privacy community must demonstrate and educate the federal agencies on these technological innovations.

While they develop these statistical disclosure control methods and communication materials, they should engage with underrepresented communities with culturally relevant outreach. As we learned with the Navajo Nation example in Chapter 1, Section 1.4, underrepresented communities must be a part of the conversation to assess if the societal benefits outweigh the privacy risks. If these communities agree to their personal data being used, they should be included at every step of the data collection and dissemination process to ensure accurate conclusions are made.

7.4 WHAT DATA PRIVACY RESOURCES ARE NEEDED?

By the time I graduated with my PhD, the data privacy community started creating more technical papers, blogs, videos,[4] and general data privacy communication outreach. This increase in communication materials will fortunately result in fewer PhD students stalking professors' web pages, and generate a better understanding of data privacy methods. Yet, most of these materials still struggle to explain data privacy concepts, especially differential privacy, for a *non-technical audience*. One of the original creators of differential privacy wrote a primer with his colleagues to help explain differential privacy for lawyers. When I talked to him about the primer, he told me that the lawyers said, "This primer is still too mathy."

This issue emphasizes the importance of defining, "Who is our non-technical audience?" The audience could be a friend, a biologist, a public policymaker, or a lawyer. Who we decide as our lay audience should motivate what examples we use, what data privacy areas we discuss, or what assumptions we should make about what the audience knows. This reasoning is why I stated in the preface that I envisioned some of the audience as my family who never graduated college, but were interested in my research. In general, we need to create several communication materials that address audiences with diverse technical and non-technical backgrounds that are interested in learning more about data privacy. Currently, the materials are limited.

Another challenge is having enough computational tools to implement the various statistical disclosure control methods. While researchers should understand the basics of statistical disclosure control methods, they do not necessarily need to thoroughly understand them. In other words, we do not need to know how to build a car in order to drive it. In this analogy, researchers are mechanics rather than car manufacturing technicians. The researchers and data users need the data privacy experts to build the basic parts for them to modify the statistical disclosure control methods for their particular data application.

---

[4]I highly recommend the "Protecting Privacy with Math" YouTube video by Minute Physics, one of my favorite physics themed YouTube Channels.

When I first started researching data privacy, I had coded several computational programs that I titled "Claire's frequently used code." At the time, there was little to no open-source[5] code on applying data privacy methods, especially differentially private ones. I wrote the code from scratch for most of my doctoral research.

Now, there are more data privacy researchers developing computational tools through several initiatives. In 2019, Harvard University and Microsoft launched OpenDP as a community to "...build a trustworthy suite of open-source tools for enabling privacy-protective analysis of sensitive personal data, focused on a library of algorithms for generating differentially private statistical releases." Microsoft also developed a computational toolkit of differentially private methods, called SmartNoise, where the automated validation server project is using and testing the code. This increase in computational tool development has not yet met the current demand for more and better computational resources.

If there are not enough people working on these problems, what about teaching the next generation of data privacy researchers? Most higher education institutions do not provide data privacy courses. If they are taught, professors usually teach them at the graduate level in computer science departments. The computer science field does contribute the most to data privacy research, but there are many other perspectives, such as statistics, economics, social science, and public policy. These other academic departments should either create data privacy course or incorporate data privacy into existing courses.

Data privacy courses should also be required at the undergraduate level and not be an optional elective for graduate students—regardless if the student wants to pursue data privacy as a career. Data science[6] and statistics[7] has significantly grown as a major and occupation in the last decade. The Bureau of Labor Statistics predicts software and analyst jobs to remain in-demand for another decade. Perspective students in these fields should understand data ethics and data privacy along with the other technical skills they

---

[5]Code that is made freely available and can be redistributed and modified.

[6]Data science is a newer scientific field that is interdisciplinary in nature due to using methods from statistics, computer science, and other fields to gain insight from data.

[7]The Bureau of Labor Statistics ranked Statisticians as the fifth fastest growing occupation in the United States. Wind turbine service technicians, nurse practitioners, solar photovoltaic installers, and occupational therapy assistants ranked higher. Ranks were last updated September 1, 2020 on https://www.bls.gov/ooh/fastest-growing.htm.

learn to avoid causing more harm than good with the data they analyze.

For instance, Amazon started a project in 2014 to automate the resume review process to select the top candidates for any position at Amazon. By the next year, the engineers on the project realized the algorithm was biased towards men for software developer jobs. The algorithm had this bias because the tech industry is predominantly male and they used resumes from the past ten years that were submitted to Amazon. In other words, the resume algorithm taught itself that Amazon preferred male candidates over female candidates by capturing keywords and phrases that were more common with one gender over another. News of this project came out late in 2018 and brought up issues of algorithmic fairness and discrimination [12].

This example is one of many other examples that demonstrate why we should better understand how we collect and analyze data to avoid introducing biases.[8] In Chapter 5, the Unacast example taught us we could be missing "true" social distancing patterns when the data are aggregated.

These situations remind us why we need educational materials to avoid creating biases in our algorithms and technologies. We cannot blindly trust the collected or analyzed data will report the specific insight we want. To quote my postdoctoral advisor,[9] "We should be focused on data-*informed* decision making instead of data-driven. We still need that human element."

These issues I covered so far lead to another problem that returns to a question posed in Chapter 4, Section 4.5. Why were there so many incorrect materials? From Chapter 4, we learned that differential privacy has an unintuitive definition, which is *still* causing many misunderstandings. The difficult definition makes creating well-thought-out and well executed educational materials challenging.

However, the problems with differential privacy does not explain why there are also few communication materials for the traditional statistical disclosure control methods, which has an intu-

---

[8]Check out *Weapons of Math Destruction: How Big Data Increases Inequality and Threatens Democracy* by Cathy O'Neil and *Data Feminism* by Catherine D'Ignazio and Lauren Klein to learn more about this area of data ethics.

[9]Dr. Joanne Wendelberger was my postdoctoral advisor at Los Alamos National Laboratory. She was a former doctoral student of Dr. George Box, a famous statistician who is often credited with the quote, "All models are wrong, but some are useful."

itive privacy loss definition. Despite the demand and importance, not many data privacy experts are interested in developing better communication materials or are in positions to do so. Even for me, I happen to be writing this book, because I randomly commented to the publisher,

> *I personally would like to write a book on "everything you need to know about data privacy and confidentiality methods" that is intended for a general audience. There is currently no such book or even a data privacy and confidentiality book that is easily accessible for those outside of the field. However, I am uncertain if CRC press or other publishing companies would be even interested.*

To my surprise, the publisher was interested and encouraged me to submit my own book proposal. But, my shock was unfounded. As we now know, few resources exist to help those outside the data privacy field understand it.

Do I believe that this book will solve our communication problems? Far from it. This book is but one drop in a large bucket of educational materials we need. As mentioned before, there are several perspectives within the data privacy community, which requires written and computational resources geared towards each of those perspectives and their data privacy needs. But, again...we must start somewhere.

Some within the data privacy community are actively working and conversing with others, such as federal public policymakers, to create accessible communication materials on the trade-offs between data utility and data privacy. These resources will be essential to have a more constructive public dialogue on how to best set this balance.

If we want to make a lasting impact on our society, we must elevate the debate by empowering policymakers to make more data-informed and evidence-based public policy decisions while ensuring and maintaining public trust with those whose data we use.

Glossary

A list of definitions from the book in alphabetical order.

administrative data
> are collected for the administration of an organization or program by entities, such as government agencies as they provide services, companies to track orders, and universities to record registered students [28]. In this book, the term is used to mean administrative data collected the Federal Government, e.g., tax records.

bottomcoding or topcoding
> a statistical disclosure control method that removes information that are below a selected lower bound or above a selected upper bound, respectively.

categorical data
> data that take on discrete values

categorical thresholding or generalizing values
> a statistical disclosure control method that groups values into broader categories.

continuous data
> data that take on any value within an interval or range

cookies
> small pieces of data generated when a person visits a website and that information is stored by their web browser. Although cookies do not contain any information that identifies the person, the information generated by cookies could be linked to other sources or be used to track people as they navigate through different websites.

data confidentiality
> refers to how the data privacy community protect participants' information in the data, such as who should have access to the sensitive data under what restrictions.

data intruder or adversary
> someone or entity that tries to disclosure sensitive information from confidential data.

data practitioners or data users
> are analysts and users of the the publicly released version of the confidential data. They often help the data stewards with applying data privacy methods on the confidential data from the perspective of those who will ultimately use the data.

data privacy
> refers to the amount of personal information individuals allow others to access about themselves [27].

data stewards, data curators, or data maintainers
> are individuals or institutions that are responsible for the collection and storage of the confidential data.

data swapping
> a statistical disclosure control method of switching observations that have similar variable characteristics.

differential privacy
> a mathematical definition or condition that quantifies privacy-loss when releasing information from a confidential dataset by linking the potential for privacy loss to how much the estimate for a unique statistic (or query) from the underlying confidential data changes with the absence or presence of any individual record that could possibly be in the dataset.

disaggregated data
> data that has been divided into detailed sub-parts.

disclosure risk
> a risk that a data intruder can use on publicly release data to obtain sensitive information of the records. There are generally three types of disclosure risks:
>
> **attribute:** a data intruder determines more accurate or new information about a record or group of records based on the structure or features of the released data.
>
> **direct:** information released that explicitly identifies individuals.

identity: a data intruder associates a record from an external data set with a specific observation in the released data.

indirect: information released that may be used in conjunction with other information to identify individuals.

inferential: a data intruder infers information about a record with high confidence based on the statistical properties of the released data.

end user
a common term in product development that refers to the people who will ultimately use the product.

establishment
a single economic unit, such as a mine, a farm, a factory, or a store. Establishments are typically at one physical location and are different from a firm, or a company, which is a business and may consist of one or more establishments.

IP or Internet Protocol address
an identification number that is assigned to every device that connects to a computer network.

longitudinal data
data that have repeated observations of the same person or record over time.

machine-readable
data in a format that can be easily processed by a computer without human intervention while ensuring no semantic meaning is lost—direct definition from the OPEN Government Act.

metadata
structural or descriptive information about data such as content, format, source, rights, accuracy, provenance, frequency, periodicity, granularity, publisher or responsible party, contact information, method of collection, and other descriptions—direct definition from the OPEN Government Act.

microlevel data
data at the individual or record level.

microsimulation model
> a model common in public policy that first estimate a baseline from current conditions in the United States and then calculate the counterfactual or alternative estimation based on the proposed policy program change.

noise infusion or sanitization
> a statistical disclosure control method of adding or subtracting random values based on a probability distribution.

privacy experts or privacy researchers
> are individuals who specialize in developing data privacy methods.

record linkage attack
> a data intruder combining one or more external data sources to re-identify individuals in publicly released data.

researchers
> are those who specialize in another area such as economics and are connected in the data privacy community, because of their need to access the confidential data.

rounding
> a statistical disclosure control method of changing the values based on certain criteria that shortens the value.

sampling
> a statistical disclosure control method of randomly selecting a subsample of the original confidential data.

statistical disclosure control or limitation
> methods of data privacy and confidentiality to safely release data publicly.

summary statistics
> a collection of statistics that typically include mean, median, minimum, maximum, and standard deviation.

suppression
> a statistical disclosure control method that removes information from the data based on certain criteria.

synthetic data
> a technique that aims to generate data with pseudo or "fake" records that are statistically representative of the original data.

tabular statistics
> count and frequency statistics.

Bibliography

[1] Ahmet Aktay, Shailesh Bavadekar, Gwen Cossoul, John Davis, Damien Desfontaines, Alex Fabrikant, Evgeniy Gabrilovich, Krishna Gadepalli, Bryant Gipson, Miguel Guevara, Chaitanya Kamath, Mansi Kansal, Ali Lange, Chinmoy Mandayam, Andrew Oplinger, Christopher Pluntke, Thomas Roessler, Arran Schlosberg, Tomer Shekel, Swapnil Vispute, Mia Vu, Gregory Wellenius, Brian Williams, and Royce J Wilson. Google covid-19 community mobility reports: Anonymization process description (version 1.0). *arXiv preprint arXiv:2004.04145*, 2020.

[2] Margo J Anderson. *The American Census: A Social History.* Yale University Press, 2015.

[3] Per Block, Marion Hoffman, Isabel J Raabe, Jennifer Beam Dowd, Charles Rahal, Ridhi Kashyap, and Melinda C Mills. Social network-based distancing strategies to flatten the COVID-19 curve in a post-lockdown world. *Nature Human Behaviour*, 4(6):588–596, 2020.

[4] Seth Borenstein. Potential privacy lapse found in Americans' 2010 Census data. *Retrieved from Associated Press News. https://apnews.com/article/aba8e57c14504 7b5bab11b62baaa 7f7a*, 2019.

[5] Claire McKay Bowen. Will the Census's Data Privacy Efforts Erase Rural America? *Retrieved from Urban Institute. https://www.urban.org/urban-wire/will-censuss-data-privacy-efforts-erase-rural-america*, 2020.

[6] Claire McKay Bowen, Victoria Bryant, Leonard Burman, Surachai Khitatrakun, Robert McClelland, Philip Stallworth, Kyle Ueyama, and Aaron R Williams. A synthetic supplemental public use file of low-income information return data:

methodology, utility, and privacy implications. In *International Conference on Privacy in Statistical Databases*, pages 257–270. Springer, 2020.

[7] Claire McKay Bowen, Ajjit Narayanan, and Corianne Payton Scally. Using Differential Privacy to Advance Rural Economic Development: Applying Data Privacy and Confidentiality Methods to Industry Employment Data. *Urban Institute*, 2021.

[8] Carole Cadwalladr and Emma Graham-Harrison. Revealed: 50 million Facebook profiles harvested for Cambridge Analytica in major data breach. *Retrieved from The Guardian. https://www.theguardian.com/news/2018/mar/17/cambridge-analytica-facebook-influence-us-election*, 17:22, 2018.

[9] James Cilke. The Case of the Missing Strangers: What we know and don't know about non-filers. In *107th Annual Conference of the National Tax Association, Santa Fe, New Mexico*. JSTOR, 2014.

[10] Jill Cowan and Matthew Bloch. In Los Angeles, the Virus Is Pummeling Those Who Can Least Afford to Fall Ill. *Retrieved from New York Times sec. The Coronavirus Outbreak. https://www.nytimes.com/interactive/2021/01/29/us/los-angeles-county-covid-rates.html*, 2021.

[11] Joseph Cox. I Gave a Bounty Hunter $300. Then He Located Our Phone. *Retrieved from Motherboard. https://motherboard.vice.com/en us/article/nepxbz/i-gave-a-bounty-hunter-300-dollars-located-phone-microbilt-zumigo-tmobile*, 2019.

[12] Jeffrey Dastin. Amazon scraps secret AI recruiting tool that showed bias against women. *San Fransico, CA: Reuters. Retrieved on October*, 9:2018, 2018.

[13] Charles Duhigg. How companies learn your secrets. *The New York Times*, 16(2):1–16, 2012.

[14] Diane H Felmlee and Derek A Kreager. The Invisible Contours of Online Dating Communities: A Social Network Perspective. *Journal of Social Structure*, 18:0_1–27, 2017.

[15] Christine Fernando and Cheyanne Mumphry. Racism targets Asian food business during COVID-19. *Retrieved from Associated Press.* https://apnews.com/article/donald-trump-race-and-ethnicity-pandemics-wuhan-animals-4d25738ab49597d0de1517383a9108d2, 2020.

[16] Centers for Disease Control, Prevention, et al. Health equity considerations and racial and ethnic minority groups. *Centers for Disease Control and Prevention, Atlanta, GA*, 2020.

[17] Ellen Galantucci. Disclosure Control Practices on BLS Administrative Data. In *113th Annual Conference on Taxation.* NTA, 2020.

[18] Suzanne Gamboa. Coronavirus reported in over half of Latino meat, poultry workers in 21 states, CDC says. *Retrieved from NBC News.* https://www.nbcnews.com/news/latino/coronavirus-reported-over-half-latino-meat-poultry-workers-21-states-n1233192, 2020.

[19] GDPR.EU. GDPR Small Business Survey: Insights from European small business leaders one year into the General Data Protection Regulation. *University of Melbourne*, 2019.

[20] Ross Goldstein, Michael E Woolley, Laura M Stapleton, Daniel Bonnéry, Mark Lachowicz, Terry V Shaw, Angela K Henneberger, Tessa L Johnson, and Yi Feng. Expanding mlds data access and research capacity with synthetic data sets. 2020.

[21] Christine Greenhow and Emilia Askari. Learning and teaching with social network sites: A decade of research in K-12 related education. *Education and information technologies*, 22(2):623–645, 2017.

[22] Heather Hansman. How a Lack of Water Fueled COVID-19 in Navajo Nation. *Outside sec. Environment.* https://www.outsideonline.com/2413938/navajo-nation-coronavirus-spread-water-rights, 2020.

[23] Nick Hart and Nancy Potok. Modernizing US Data Infrastructure: Design Considerations for Implementing a National Secure Data Service to Improve Statistics and Evidence Building. *Data Foundation*, 2020.

[24] Drew Harwell. Colleges are turning students' phones into surveillance machines, tracking the locations of hundreds of thousands. *Retrieved from The Washington Post.* *https://www.washingtonpost.com/technology/2019/12/24/* *colleges-are-turning-students-phones-into-surveillance-* *machines-tracking-locations-hundreds-thousands/*, 24, 2019.

[25] History Staff Jason Gauthier. Privacy and Confidentiality - History - U.S. Census Bureau. *https://www.census.gov/history*, 2020.

[26] Suzanne Rowan Kelleher. Scoreboard: These States Are Nailing Social Distancing. *Retrieved from Forbes.* *https://www.forbes.com/sites/suzannerowankelleher/2020/* *03/29/scoreboard-these-states-are-nailing-this-social-* *distancing-thing/*, 2020.

[27] Sallie Ann Keller, Stephanie Shipp, and Aaron Schroeder. Does Big Data Change the Privacy Landscape? A Review of the Issues. *Annual Review of Statistics and Its Application,* 3(1):161–180, 2016.

[28] Sallie Ann Keller, Stephanie S Shipp, Aaron D Schroeder, and Gizem Korkmaz. Doing data science: A framework and case study. *Harvard Data Science Review 2(1), 2020.* *https://hdsr.mitpress.mit.edu/pub/hnptx6lq/release/8?reading* *Collection=b97b8a1b*

[29] Cameron F. Kerry and Caitlin Chin. By passing Proposition 24, California voters up the ante on federal privacy law. *Retrieved from Brookings Institute.* *https://www.brookings.edu/blog/techtank/2020/11/17/by-* *passing-proposition-24-california-voters-up-the-ante-on-* *federal-privacy-law/*, 2020.

[30] Cameron F. Kerry, John B. Morris, Caitlin Chin, and Nicol Turner Lee. Bridging the gaps: A path forward to federal privacy legislation. *Retrieved from Brookings Institute. https://www.brookings.edu/research/bridging-* *the-gaps-a-path-forward-to-federal-privacy-legislation/*, 2020.

[31] Danica Kirka. Amid confusion, EU data privacy law goes into effect. *Retrieved from Associate Press.* *https://apnews.com/article/3b6945f9f5794d87bb5c78bb093f7* *24a*, 2018.

[32] Gina Kolata. Your Data were 'Anonymized'? These Scientists Can Still Identify You. *Retrieved from New York Times sec. Health. https://www.nytimes.com/2019/07/23/health/data-privacy-protection.html*, 2019.

[33] Danielle Kwon. Confronting Racism and Supporting Asian American Communities in the Wake of COVID-19. *Retrieved from Urban Institute. https://www.urban.org/urban-wire/confronting-racism-and-supporting-asian-american-communities-wake-covid-19*, 2020.

[34] Amy Lauger, Billy Wisniewski, and Laura McKenna. Disclosure avoidance techniques at the US Census Bureau: current practices and research. *Center for Disclosure Avoidance Research, US Census Bureau*, 2014.

[35] Roslyn Layton. The 10 Problems of the GDPR: The US Can Learn from the EU's Mistakes and Leapfrog Its Policy. *AEI Paper & Studies*, page 1, 2019.

[36] Casey Leins. Which States Are Best at Social Distancing. *Retrieved from U.S. News. https://www.usnews.com/news/best-states/articles/2020-03-27/these-states-are-best-at-social-distancing-during-the-coronavirus*, 2020.

[37] Ramona-Diana Leon, Raúl Rodríguez-Rodríguez, Pedro Gómez-Gasquet, and Josefa Mula. Social network analysis: A tool for evaluating and predicting future knowledge flows from an insurance organization. *Technological Forecasting and Social Change*, 114:103–118, 2017.

[38] Fang Liu, Evercita Eugenio, Ick Hoon Jin, and Claire Bowen. Differentially private generation of social networks via exponential random graph models. In *2020 IEEE 44th Annual Computers, Software, and Applications Conference (COMPSAC)*, pages 1695–1700. IEEE, 2020.

[39] Mary Madden. Privacy, security, and digital inequality. *Data & Society*, 2017.

[40] Rowland Manthorpe. Wetherspoons Just Deleted Its Entire Customer Email Database—On Purpose. *Retrieved from Wired. https://www. wired. co. uk/article/wetherspoons-email-database-gdpr*, 2017.

[41] Harry Maugans. 5 ways to finally fix data privacy in America. *Retrieved from VentureBeat. https://venturebeat.com/2021/01/29/5-ways-to-finally-fix-data-privacy-in-america/, 2021.*

[42] Aletha Maybank. Why racial and ethnic data on COVID-19's impact is badly needed. *American Medical Association*, 2020.

[43] Sheena McKenzie. Facebook's Mark Zuckerberg says sorry in full-page newspaper ads. *Retrieved from CNN. https://www.cnn.com/2018/03/25/europe/facebook-zuckerberg-cambridge-analytica-sorry-ads-newspapers-intl/index.html, 2019.*

[44] Gordon Mermin, Len Burman, and Frank Sammartino. An Analysis of Senator Bernie Sanders's Tax and Transfer Proposals. *Tax Policy Center*, 2, 2016. *https://www.urban.org/research/publication/analysis-senator-bernie-sanderss-tax-and-transfer-proposals*

[45] Gordon B Mermin, Janet Holtzblatt, Surachai Khitatrakun, Chenxi Lu, Thornton Matheson, and Jeffrey Rohaly. An Updated Analysis of Former Vice President Biden's Tax Proposals. *Tax Policy Center*, 2020.

[46] Commission on Evidence-Based Policymaking. The promise of evidence-based policymaking: Report of the Commission on Evidence-Based Policymaking, 2017.

[47] Carly Page. H&M Hit with Record-Breaking GDPR Fine Over Illegal Employee Surveillance. *Retrieved from Forbes. https://www.forbes.com/sites/carlypage/2020/10/02/hm-hit-with-record-breaking-gdpr-fine-over-illegal-employee-surveillance/?sh=2083688715ab, 2020.*

[48] Luc Rocher, Julien M Hendrickx, and Yves-Alexandre De Montjoye. Estimating the success of re-identifications in incomplete datasets using generative models. *Nature communications*, 10(1):1–9, 2019.

[49] Ryan Rogers, Subbu Subramaniam, Sean Peng, David Durfee, Seunghyun Lee, Santosh Kumar Kancha, Shraddha Sahay, and Parvez Ahammad. Linkedin's audience engagements api: A privacy preserving data analytics system at scale. *arXiv preprint arXiv:2002.05839*, 2020.

[50] Tony Romm, Elizabeth Dwoskin, and Craig Timberg. US Government, tech industry discussing ways to use smartphone location data to combat coronavirus. *Retrieved from Washington Post. https://www. washingtonpost. com/technology/2020/03/17/white-house-location-data-coronavirus/. Accessed*, 30, 2020.

[51] Adam Satariano. Google is fined \$57 million under Europe's data privacy law. *Retrieved from The New York Times. https://www.nytimes.com/2019/01/21/technology/google-europe-gdpr-fine.html*, 21, 2019.

[52] Corianne Payton Scally, Kathryn L.S. Pettit, and Olivia Arena. 500 Cities Project Local Data for Better Health. *Urban Institute*, 2017.

[53] Mark Scott, Laurens Cerulus, and Laura Kayali. Six Months In, Europe's Privacy Revolution Favors Google, Facebook. *Retrieved from POLITICO EU. https://www.politico.eu/article/gdpr-facebook-google-privacy-data-6-months-in-europes-privacy-revolution-favors-google-facebook/*, 2018.

[54] Natalie Shlomo. Statistical disclosure limitation: New directions and challenges. *Journal of Privacy and Confidentiality*, 8(1), 2018. *https://journalprivacyconfidentiality.org/index.php/jpc/ article/view/684*

[55] Hollie Silverman, Konstantin Toropin, Sara Sidner, and Leslie Perrot. Navajo Nation surpasses New York state for the highest COVID-19 infection rate in the US. *Retrieved from CNN. https://www.cnn.com/2020/05/18/us/navajo-nation-infection-rate-trnd*, 2020.

[56] Oliver Smith. The GDPR Racket: Who's Making Money From This \$9 bn Business Shakedown. *Retrived from Forbes: https://www.forbes.com/sites/oliversmith/2018/05/02/the-gdpr-racket-whos-making-money-from-this-9bn-business-shakedown/?sh=5ee283a234a2*, 2018.

[57] Joshua Snoke and Claire McKay Bowen. How Statisticians Should Grapple with Privacy in a Changing Data Landscape. *CHANCE*, 33(4):6–13, 2020.

[58] Jeff South. More than 1,000 us news sites are still unavailable in Europe, two months after GDPR took effect. *NiemanLab, August,* 7, 2018.

[59] Matthias Templ, Bernhard Meindl, Alexander Kowarik, et al. Introduction to statistical disclosure control (SDC). *Project: Relative to the testing of SDC algorithms and provision of practical SDC, data analysis OG*, 2014.

[60] Stuart A Thompson and Charlie Warzel. Opinion— Twelve million phones, one dataset, zero privacy. *Retrieved from New York Times sec. Opinion. https://www.nytimes. com/interactive/2019/12/19/opinion/location-tracking-cell-phone.html*, 2019.

[61] Joe Tidy. British Airways fined $20m over data breach. *Retrieved from BBC News. https://www.bbc.com/news/technology-54568784*, 2020.

[62] Yanxin Wang, Jian Li, Xi Zhao, Gengzhong Feng, and Xin Robert Luo. Using Mobile Phone Data for Emergency Management: a Systematic Literature Review. *Information Systems Frontiers*, pages 1–21, 2020.

[63] Darrell M. West. How employers use technology to surveil employees. *Retrieved from Brookings Institute. https://www.brookings.edu/blog/techtank/2021/01/05/how-employers-use-technology-to-surveil-employees/*, 2021.

[64] Gus Wezerek and David Van Riper. Changes to the census could make small towns disappear. *Retrieved from New York Times sec. Opinion. https://www.nytimes.com/interactive/2020/02/06/opinion/census-algorithm-privacy.html*, 2020.

Index